THE DISCOVERY
OF THE UNIVERSE

THE DISCOVERY OF THE UNIVERSE

A HISTORY OF ASTRONOMY AND OBSERVATORIES

CAROLYN COLLINS PETERSEN

AMBERLEY

To all the stargazers, past and present.

Page 1: The Medicine Wheel in the Bighorn National Forest, situated in Wyoming in the western United States, is a sacred site and observatory. Some of its stones appear to align with specific rising and setting times of the Sun on the solstices and rising and setting points of various stars. (Courtesy: United States Forest Service)

First published 2019

Amberley Publishing
The Hill, Stroud
Gloucestershire, GL5 4EP

www.amberley-books.com

Copyright © Carolyn Collins Petersen, 2019

The right of Carolyn Collins Petersen to be identified as the Author of this work has been asserted in accordance with the Copyrights, Designs and Patents Act 1988.

ISBN 978 1 4456 8413 0 (hardback)
ISBN 978 1 4456 8414 7 (ebook)

British Library Cataloguing in Publication Data. A catalogue record for this book is available from the British Library.

Typesetting by Aura Technology and Software Services, India. Printed in the UK.

CONTENTS

'Not in vain do we watch the setting and rising of the stars.'
Inscription on a gold medal given to astronomer Maria
Mitchell for her discovery of a telescopic comet.

There is stardust in your veins. We are literally, ultimately children of the stars.
Dame Susan Jocelyn Bell Burnell

No one regards what is before his feet; we all gaze at the stars.
Quintus Ennius (239–169 B.C.)

If all mankind could look through that telescope, it would change the world.
Philanthropist Griffith J. Griffith, founder of
the publicly accessible Griffith Observatory,
Mount Hollywood, California, USA

Cosmology brings us face to face with the deepest mysteries, questions that were once treated only in religion and myth.
Carl Sagan

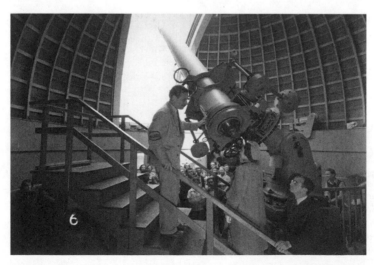

Since its opening in 1935, Griffith Observatory has made its telescopes available for public viewing, free of charge, on clear nights. This image shows observatory guides getting ready for a night of observing during the facility's opening in 1935. Courtesy: Griffith Observatory.

PREFACE

I have always been interested in observatories. As a small child growing up in a university town, I was aware the campus had an observatory. That domed building is still there, up on a hill overlooking much of the city and a nearby planetarium. But there were also other places in the area where radio astronomy installations and radar antennae connected to the cosmos, too. Once, when playing in a field near our house, I came across an instrument package and its radiosonde balloon. My father called the number on the box and some scientists from a nearby lab came out to pick it up. They told us it was being used to study the upper atmosphere of our planet and Earth's magnetic field. That was the first time I realised we could detect such things. At the tender age of six, I had seen a mini-observatory with my own eyes and picked it up off the ground.

That discovery was only a small part of the impetus for this book. Other people certainly influenced my interest in the stars – among them my dad, my boyfriend (who became my husband), my mother (a teacher who encouraged me to read) and – oddly enough – Walt Disney, who used stars as a motif in many of his shows. I also credit my astronomy teachers from college onwards. They pointed the way, I looked up, and fell in love with the sky all over again. The more I learned, the more I wanted to learn. And, when it came time for me to use an observatory for the first time, I was awestruck.

To truly tell the story of all the observatories across all of human history would require many times more pages than I have here. So, I've worked on a straightforward look at observatories, humanity's use of them, and what we've learned from these cosmos mariners. This book is really intended as a very general survey for readers who are interested in observatories but don't have an extensive knowledge of them. There are those whose interest will be piqued to read further, to dig into technical details of mirrors and spectrographs and detectors and so on. For them, I've provided further research directions in the Appendix. That material will guide readers to more technically oriented discussions. As technology is ever-changing, this is the best way to point the reader toward more detailed information about what's 'hot and new' in observatory instrumentation.

ACKNOWLEDGEMENTS

This story wouldn't be possible without access to the vast amounts of information the world's observatories provide via their web pages and other publications. In particular, I'm grateful to the public outreach offices at the world's observatories and space agencies for the images and information they make available.

Special thanks go to the various astronomers I've talked with over the years about various aspects of observatories, including Dr John C. Brandt (who provided unique insights about observatories while I was his grad student), Dr Heidi Hammel, Dr Edwin C. Krupp (whom I worked with on a major project at Griffith Observatory and whose knowledge of ancient observing sites is verging on omniscient) and Dr Erin MacDonald (who filled me in on the intricacies of gravitational wave detectors).

Many thanks to Stuart Atkinson, Dr Kenneth G. Carpenter, Keith Gleason (formerly of the University of Colorado's Sommers-Bausch Observatory), and Richard Wright (amateur astronomer and astrophotographer extraordinaire) for reading parts of this book and keeping me on the straight and narrow. Any errors or omissions that remain are mine. Special thanks to Mark C. Petersen for providing loving support during the year this book was making its way from my brain to these pages.

INTRODUCTION

We've seen them on the news, or in a passing landscape – dome-topped buildings on hilltops, radio arrays stretching out across the countryside, and otherwise mysterious-looking structures scattered across university campuses and people's backyards. Most are located on Earth, but a relative few are in space. What are these places? They're observatories, and they have been around in one shape or another since humanity's earliest history. For astronomers, they're places of work, the tools of cosmic exploration. This book is a general look at observatories, their histories, and the way they are being used today.

Observatories are humanity's portals to the universe. They are, in all but the most literal sense, multi-wavelength mariners that guide us across the gulfs of space and time, taking our eyes and ears and minds on voyages of discovery. The telescopes they contain are time machines, revealing distant objects as they once existed and showing us what familiar objects such as the Sun will look like in the distant future. Because astronomy is our oldest science, observatories have played a big role in our ongoing discovery of the universe. So, what are these places? What do they contain? How do they work? That's the story we explore together in this book.

The story of astronomy is filled with tales of lone heroes peering at the skies and studying the objects overhead. Early in history, intrepid observers started simply enough, with naked-eye observations. Their observatories were windswept hilltops, mountaintops, massive pyramids, and stone circles. Lacking technology, all they could do was carefully study and map the stars, chart the planets, and note anything out of the ordinary. For centuries, this naked-eye observing was how astronomy was done. This changed with the invention of the telescope in the early 1600s. Those first instruments heralded the birth of a new era in skywatching — a magnified one, where a simple tube and some precisely ground optical glass cast humanity's gaze across great distances. From that time onwards, the term 'observatory' took on new meaning: it came to mean the place, often a domed building, where telescopes are used to look at the sky. With a few minor changes to accommodate modern instruments and sophisticated arrays, it's still how many observatories function today. However, as we'll see in this book, not all of them look like observatories. So the definition may need changing again.

The story of the observatory is not just a story of a piece of equipment, however. It's also the tale of people with a relentless passion for the sky, who pushed themselves and their instruments to look farther with sharper optics. In time, they added cameras to their observing toolkit, as well as other instruments and detectors. With more complexity of instrumentation, the practice of Big Astronomy moved from the realm of the single observer at the eyepiece to extended teams working together around the planet.

Today, astronomy is done largely by those teams, comprising astronomers who gather terabytes of data into vast treasure troves of information about objects and processes in the universe. They write and use computer code to analyse the data and eventually publish their findings. Big Astronomy's instruments may well be telescopes as we traditionally think of them – assemblages of metal and mirrors and instruments. But modern astronomers also use radio antennas, orbiting spacecraft, and, most recently, gravitational wave detectors set half a continent apart here on Earth. Observatories today are still cosmos mariners, taking our vision and our imaginations sailing across the light-years. They reveal not just distant objects but an understanding of the universe — and, ultimately, our place in the cosmos.

Learning the Sky as a Student

My first professional experiences with telescopes were at two very different facilities. As a graduate student at the University of Colorado back in the 1990s, I had access to a well-equipped amateur facility not far from where I currently live in the Rocky Mountains of Colorado. From there, I watched comets with a genial gentleman named Gary Emerson. He often made his telescopes available to me at short notice. From him, I learned the rudiments of observing and imaging.

Based on that work, I next observed using the University of Hawaii's 2.3-metre telescope on the summit of Mauna Kea to observe the (at the time) newly discovered Comet Hale-Bopp. I'll never forget the eight nights I spent on top of the mountain with a very interesting crew of people, led by astronomer Dr Karen Meech, her then-assistant Dr Olivier Hainaut, and their graduate student James 'Gerbs' Bauer. It was, for me, the pinnacle of observing, in more ways than one. At 4,207 metres (13,803 feet) above sea level, Mauna Kea provides some of the darkest, driest skies available on the planet. It's perfectly suited for observations of anything from comets to distant galaxies. On more than one occasion during the observing run I stepped outside to look upwards at the sky. Each time, I was impressed with the immensity of a universe that could hold both me and the comet I wanted to observe. At one point, I felt as if I could simply leap off the catwalk and up to the stars, free to explore as I wished. I'm sure some of my reaction was due to oxygen deprivation, but a larger part of it was seeing the very dark sky overhead set against the backdrop of the otherworldly surroundings atop the mountain.

Also as part of my work, I had the chance to be part of an instrument team for the Hubble Space Telescope (HST). For those who aren't familiar with this venerable telescope, on orbit for more than two decades, it's a workhorse 3.4-metre mirror mounted in a tube the size of a school bus, orbiting the planet once every 90 minutes. From its orbital perch at 547 kilometres (339 miles), Hubble has brought the universe ever closer to us. During my time as a graduate student, I got to know it quite well, and co-wrote one of the first books about HST science ever published.

These days, I work with other observatories from time to time, and each one has a unique flavour. They're all located in remote places, the better to get the most detailed views of distant places in the cosmos. Their stories are worth telling, because observatories are how humanity learned to voyage through the stars without ever leaving our planet. That's what this book is all about: introducing the reader to the stories of the stars, from the places where those tales were first uncovered.

Exploring the World of Observatories

During this journey, we look at observatories throughout human history and observe their connection to the development of the telescope. It's impossible to cover every observatory in this short space, but I've selected as many interesting ones as possible. Of course, new ones are coming online every year, building on the accomplishments of their older siblings and cousins. I hope this very general introduction serves the purpose of acquainting the reader with the vast array of observatories that are out there, turning out good science 24 hours a day, seven days a week.

The book starts with a very general look at the main components of an observatory. Then, in order to understand why people use these places to observe what they do, and how they do it, we look at the astronomical science they enable. The history of observatories plays out in Chapter 3, largely focused on some aspects of the parallel development of the telescope and observatory facilities in the 17th and 18th centuries. From there, we will look at the great observatories of the 19th and 20th centuries. Finally, we will explore selected 21st-century institutions (including space-based facilities), ending with the observatories of the future. The book closes with a look at amateur observatories, public astronomy education, citizen science initiatives that capitalise on the tremendous data sets observatories are churning out, and a brief foray into the observatory in media and popular culture.

I

INTRODUCTION TO THE OBSERVATORY

We are people of the stars. When we look at the sky, we are looking at our ancient home. That's because we are, as the late astronomer Carl Sagan often told us, made of star stuff. So, at some level, it makes sense for us to look at the stars and try to understand them. Of course, it has taken generations of skygazers to gain the knowledge of the sky humanity shares today. Every age has its observers who make amazing contributions using the tools at their disposal. In a very real sense, then, we are descended from astronomers. We owe our existence, our survival, our concepts of time, space, and the objects in the universe to generations of people who have gazed at the sky.

The invention of the telescope made pivotal changes in astronomy that allowed humans to discover the universe. Before the time of Galileo, people had only their eyes with which to study celestial objects. The sky was a dome arching overhead, and all its planets and starry backdrop were too distant to be probed. The telescope magnified our view and brought the sky closer. It provided humans with a way to measure and study the sky.

We changed from a species which simply 'used' the sky to one that explored it. Discoveries made using the telescope, and variations on it, continue today under domes, in backyards and dark-sky sites around the planet. We have orbiting observatories and robotic probes that explore other worlds using purpose-built instruments to send back detailed information about those distant places. In recent years, some facilities look more like giant swimming pools or are located far under the ocean surface. In the hands of astronomers, planetary scientists, solar physicists and others, the observatory, regardless of what it resembles, remains our cosmos mariner – the metaphorical ship taking us on voyages of discoveries throughout the universe.

Who Are the Astronomers?

To understand observatories and what they mean to us, it's useful to begin by looking at the people who use these tools of astronomy. In every age, they have advanced their technology and their science, benefitting today's practice of astronomy and astrophysics. Think of them as captains of our observatory ships, setting course for the planets, stars, and galaxies.

Astronomers come from all walks of life. The first ones may well have been cave-dwellers who stepped outside to marvel at the sky. The universe they were concerned with was a very local one: the Sun, Moon, and planets, all moving against a backdrop of stars. It likely didn't take them long to make connections between the motions of these objects throughout the course of a year.

Over the centuries, Chinese astrologers, Greek philosophers, Persian astronomers, and Polynesian navigators were among the many people who codified that connection between the sky, the march of time, and finding one's way around our planet. Their early work set the stage for the later development of the science of astronomy. Over the course of thousands of years, men and women have worked to chart the sky and map the stars.

Modern astronomers follow in their footsteps, on cosmic trails blazed by others who were fascinated with the sky. Today we have amazingly sophisticated instruments that take teams of scientists and technical experts to operate and collect data. We in the modern world are indebted to those who came before us and opened the sky to our eyes. They, like today's astronomers, exhibited tenacity, ingenuity, and an ability to look beyond Earth and use science to understand what they saw overhead.

Miss Mitchell and the Comet

One night in 1847, a young woman stood atop a building on the island of Nantucket in Massachusetts. She was peering through a telescope at something in the sky. It looked faint and fuzzy, just like a comet. Could it *be* a comet? That's what Miss Maria Mitchell thought to herself as she took notes and made sketches in her notebook.

Miss Mitchell was a strong-willed young hero pitting herself against the universe – and social convention. She lived at a time when women weren't expected to 'do' science. It was unfeminine, or so society said. However, Maria never cared about what society thought. She was raised to study the sky, do the math, and analyse what she'd seen, just as her father taught her to do. If it was all the same to everyone else, she'd continue her explorations and studies, thank you very much.

So, it was no surprise to find her at her favourite 'observatory' on the roof of a library. That night, she discovered a comet and it made her famous. Today it has the designation C/1847 T1, but to people of her time, it was 'Miss Mitchell's Comet' and it made Maria's name a household word. Her observatory wasn't anything like what we expect to see today, just a simple telescope and a place to set it up. But spying out a comet that night was perfectly in the spirit of discovery which began when the first humans raised their eyes to the skies and made simple records of what they saw. It continued through generations of naked-eye stargazers around the world who learned to use the stars.

An Astronomer Discovers the Universe

There's a familiar meme in astronomy circles of the 'lone hero at the telescope'. It describes a person dedicated to spending his or her time gazing at the sky, looking for something dim and distant. Certainly Miss Mitchell fits this description, and there have been many others throughout astronomy history: Galileo, the Herschels, Messier, Lowell, and Tombaugh, to name a few.

Early in the 20th century, Edwin P. Hubble made an observation that put him squarely into the category. He spent many nights at the Hooker Telescope on Mount Wilson studying the sky and taking photographic plates. He had an idea the universe might be bigger than astronomers thought. At the time, the commonly accepted idea was that the Milky Way Galaxy was the sum total of the universe. Even the great Harlow Shapley, who had made the first and most extensive survey of stars in the Milky Way to determine its size, believed our galaxy was all there was.

Still, there were these pesky things that kept turning up in the photographic plates. They had first been labelled 'spiral nebulae' in the 1800s by Herschel and others, but most astronomers decided these fuzzy spirals were likely inside our galaxy. There wasn't any empirical data to suggest otherwise.

Hubble wasn't so sure. During his nights at the Hooker eyepiece, he zeroed in on something he thought was a simple nova – a star flaring into brightness in the Andromeda 'spiral' nebula. He studied photographic plates taken earlier by Shapley and others, and realised the star was actually a Cepheid variable. This is a type of star that varies in its brightness over a known period of time. Astronomer Henrietta Leavitt had established through her studies of these strange stars that their stable and periodic brightenings and dimmings could be used to establish their distances.

So, Hubble used the period of the star to determine its distance and it turned out to be much farther away than any star thus measured in the Milky Way. That meant the Andromeda Nebula was also very far away. Hubble began measuring and studying the spectra of all the known spiral nebulae, and found them all to be very distant. It was a startling conclusion and essentially marked the discovery of the larger universe. Thanks to his work, astronomers realised not only was the universe bigger than they expected it to be, but it was expanding.

The Grad Student and the Little Green Men

One hundred and twenty years after Miss Mitchell's discovery and 44 years after Edwin Hubble showed the universe is expanding, a young woman peered skyward. She wasn't using a giant telescope on a mountain. Instead, her observatory was a rudimentary radio array. Jocelyn Bell (now Dame Jocelyn Bell Burnell) was a Ph.D. researcher in 1967 when she detected what she called a 'bit of scruff' on a chart recording of sky observations she had made. Her instrument was called the Interplanetary Scintillation Array, a radio telescope built at Mullard Radio Astronomy Observatory in Cambridge, England.

She was using that array to sweep the sky in a search for quasars. These are bright, point-like sources with very active bright central regions. The 'scruff' she found was a very regular signal recurring over and over again. It quickly got dubbed 'LGM-1' for 'Little Green Men'. Far from being aliens, however, her discovery was of objects we now know as pulsars, and – as with Miss Mitchell – she faced resistance to her work in some part due to her gender. She was skipped over for a Nobel Prize for her discovery (it went instead to her advisor). However, her experience marked two turning points: the discovery of a pulsar, and a change in the social attitudes toward women in astronomy. It's one that the field is still grappling with even in the 21st century, as more and more women enter the field to study everything from planets to those very same pulsars the young Jocelyn discovered.

Ms Bell's work was done with a detector very few laypersons would recognise as a 'telescope'. She helped build it and deploy it in the search for radio-loud objects. It looked for all the world like a fence with wires strung along the posts. Yet, this ungainly installation was good enough to provide the data behind one of the most significant scientific achievements of the 20th century. Today, Dame Jocelyn Bell Burnell is one of more than 11,000 professional astronomers around the world who constantly study the sky, make observations, and draw interesting conclusions about the universe.

A Team Approach to a Black Hole
Late in the first decade of the 21st century, a young man described an idea his team was pursing for a fantastic globe-girdling telescope that might be capable of imaging the event horizon of a black hole. It would be an amazing feat of technology and science. Black holes look, well, black. They can't be seen because the heart of a black hole is a tiny singularity stuffed full of mass, which has a strong gravitational pull. Any light entering the mass would be lost forever. Any light passing close by would have its path distorted, and thus the 'picture' of a black hole's event horizon would also similarly be distorted. That was the challenge Dr Shepherd Doeleman and his team faced when they began planning an observation of a supermassive black hole. They knew radio astronomy would give them the clearest view of the black hole region. However, to get the

clearest view of the closest region *to* the black hole, they needed a very powerful radio telescope. That's when he put together the Event Horizon consortium, uniting scientists around the world. Together, they proposed linking some of the most powerful radio telescopes on the globe into one virtual telescope. It would be focused on the black hole and the combined power of all the telescopes would be a good use of a technique called 'interferometry'. Essentially, they would all gather data as the object came into their view. The information would be combined and correlated. The result would be pumped into a mathematical algorithm that would create an 'image' of the event horizon.

This resulting Event Horizon Telescope (EHT) united people around the world in a common goal to image a black hole. It's a good example of the team approach to astronomy, which dominates the science in the 21st century. With ever-increasing data loads and international cooperation, the team approach is the best way to bring the combined intellects of talented astronomers to focus on solving a scientific problem.

Astronomers are most frequently associated with research institutions in government and university environments around the world. Some are also teaching professionals, bringing up the next generations of astronomers, observers, computer programmers and theoreticians to take advantage of the massive amounts of data churned out by modern observatories. With the advent of team structures, the age of the 'lone hero' astronomer staying up many nights to make crucial studies and discoveries is mostly long gone.

Defining 'Observatory'
Astronomical observatories have existed since prehistory. Their chief focus throughout history was to provide access to the sky, whether from a mountaintop, a mound, a pyramid, or in more recent times, from space or from facilities around the planet connected by computer. There are hundreds of ground-based observatories on Earth used mainly by professional astronomers. There are also many more amateur-type facilities, ranging from a simple telescope mounted in someone's backyard to research-grade installations maintained by dedicated amateurs and astronomy clubs.

Today, when we think of 'observatory', many people visualise those giant telescopes in buildings on mountaintops. Or they may think of the orbiting Hubble Space Telescope, or the planetary probes that are studying Mars, for example. Astronomical observatories, particularly those in research institutions, not only have some sort of light-gathering capability, but they're also well-equipped with instruments and cameras. They are constantly maintained by highly trained technicians who do everything from re-coating mirrors to installing new instruments and infrastructure, repairing technical breakdowns and calibrating equipment.

In times past, professional astronomers would travel to the observatory to do their work. Today, with the advent of high-speed internet connections and automated observatory processes, such travel is not strictly necessary. It's not as if the astronomers get to look through the eyepiece at their target and then take images or data. These days the work is done by computer. Some do choose to 'go to the mountain' and they can monitor the observation on a computer screen if they wish. However, many simply order the observations and get them approved through telescope time allocation committees. Then they wait for the data to be taken, processed, and delivered to them for study.

The largest exception to this rule comprises the amateur astronomy community members. Many of these people have built their own observatories. While they may automate and computerise some aspects of observation, they still go out to their domes and telescopes in great numbers. And, as we'll see in the last chapter, many amateurs are involved in 'professional-amateur' collaborations, lending their expertise and telescopes to ongoing observation programmes which depend on them for data.

We also call things 'observatories' that don't fit in with the 'sky-facing' or 'space-facing' definition. Many focus on terrestrial events and objects, such as the Hawaiian Volcano Observatory, or meteorological observatories that are focused on weather, and oceanographic installations that monitor the deep seas. In many places in the world, there are installations called 'observatories' which allow people in fire-prone areas to keep an eye out for wildfires. 'Observatory' is a good multi-purpose word for places where we 'watch' for things. However, for the purposes of this book, the focus is on the cosmos mariners, the places where people go to study objects in the sky. They might be in giant buildings, or they can be in small, backyard setups. Some are deep underground while others are spread out across desert landscapes. They might be single telescopes with simple cameras, or behemoths with tons of equipment attached to them, or arrays with lengthy names that make cool acronyms – and astronomers *love* acronyms, as we will see! Whatever these observatories look like or are called, they continue to transform our view of the cosmos.

Optical Telescopes

The telescope is the heart of many an observatory. An optical telescope detects visible light, which is often referred to as 'optical' wavelengths. They enter the 'tube' of the telescope, bounce off of a mirror, and are directed into an array of detectors and instruments. This is true regardless of whether an optical telescope is for professional or amateur use.

Such an instrument can be in one of three basic designs: refractor, reflector, and catadioptric (with a few variations in design). A refracting telescope uses two lenses to deliver a view of a celestial object. At one end (the one farther away from the viewer), is a large lens, called the 'objective lens'. The other end holds the lens the user looks through. It is called the 'ocular' or 'eyepiece'. They work together to deliver the sky view. The objective lens (or mirror) focuses the light from the distant object to form a tiny image of the object within the telescope itself; the ocular serves as a magnifying glass that provides an enlarged and detailed view of that image.

A reflector telescope gathers light onto a concave mirror at the bottom of the scope. This is what astronomers call the primary and it's usually ground into a parabolic shape. It's easy to remember, since it 'reflects' light, as opposed to gathering it through a lens. The first reflecting type of telescope was developed by Sir Isaac Newton.

Catadioptric telescopes combine elements of refractors and reflectors in their design. The most common type today is a variant called the Schmidt-Cassegrain that has good observing power in a compact design. Many observatory telescopes, such as Gemini in Hawaii or the orbiting Hubble Space Telescope, are variants of this type of telescope.

Newer facilities, such as the Large Synoptic Survey Telescope, funnel gathered light through a three-mirror lens optical assembly. It's then sent to a detector, camera, or other instruments for further imaging and analysis.

Most telescopes are set on mounts, which allow them to swivel around to aim at specific areas of the sky and track them as Earth rotates. This is true for all ground-based optical, infrared, radio, and submillimetre instruments. In space, telescopes move around by using specialised reaction wheels called 'gyros' (short for 'gyroscopes') that stabilise the instrument and help it swing around to point in the desired direction.

All observatories, whether they are on a mountain or in a backyard or even buried deep underground, must operate under optimum conditions. Mirrors in telescopes need to acclimate to the ambient air temperature, for example. Glass, like any other material, can expand and contract under extremes of heat and cold. Some of this can be avoided by the mirror cell's design, but most facilities also take great pains to make sure the telescope is at the proper temperature for the best viewing. The Gemini Observatory, for example, has a shutter system that is opened during observations to allow the dry mountain air to blow through, which enables the mirrors to reach a stable temperature.

Amateur observers either have their telescopes mounted in a homebuilt observatory, or they carry their instrument outside to let the mirror stabilise. If they use a CCD (charge coupled device) camera often enough it will have a heat sink and/or fan on the back to carry away excess heat from the chip. Detectors on the large telescopes also have to be cooled down to a temperature at which they are stable and unbothered by heat-induced 'noise'. Some use nitrogen gas for the purpose. In particular, infrared-sensitive detectors need to be kept as 'chilled' as possible for the best observations.

Some space-based detectors are also kept at specific temperatures (using cooling systems or heating elements) to ensure their optimum performance. Other instruments, such as neutrino detectors, are located deep underground to avoid stray cosmic rays.

Astronomy at All Wavelengths and Frequencies

All telescopes gather light, but certain types of light require different-looking instruments. Visible light is what telescopes were originally built to detect. That is the light we see with our eyes and astronomers usually refer to it as 'optical'. It's a small part of a larger spectrum that includes everything from gamma rays through to x-rays, ultraviolet, infrared, microwave, and radio frequencies. These parts of the spectrum are invisible to our eyes, and require specialized observation methods. Most astronomy throughout history has been done with 'optical light' until relatively recently.

Beyond the optical, many ground-based observatories at high altitudes and in dry climates can and do observe infrared light that is close to the optical range. It's called the 'near-infrared'. Places such as the Keck and Gemini Observatories can do observations in that range using the same mirrors they use for optical observations. Beyond what Gemini, Keck, and others can see in the infrared, astronomers need to use space based observatories. Infrared astronomy from space allows observations unhampered by Earth's atmosphere, and allows astronomers to look at very young objects as they existed at the beginning of the universe. Infrared light also passes easily through dust clouds around young stars, allowing their radiation to be detected. In addition, solar system objects can be spotted in infrared light.

In the submillimetre range, instruments used to detect those emissions look like small radio telescopes. Instruments used for radio, x-ray, and gamma-ray astronomy may not look as familiar as the optical telescope, but they still gather light we, as humans, can't see.

How the Rest of the Spectrum was Discovered

In 1800, Sir William Herschel discovered what he called a type of 'invisible' radiation, which turned out to be infrared light. It happened as he was studying the effects of different filters for his telescope. He put a prism in sunlight and measured the temperatures of the colours he could see, noting that they increased from the blue to the red part of the spectrum. When he put a thermometer just past the red part of the spectrum, where there was no visible light, he noted the temperature there was even higher. Herschel reasoned there must be another type of light that we can't see. We call it infrared today, but he called it 'calorific rays'. Further study revealed that this light could be reflected, refracted, absorbed, and transmitted, the same as visible light. His discovery eventually led to studies of other forms of light, as part of the science of spectroscopy, which dates back to the time of Isaac Newton.

Today, astronomy is a multi-wavelength discipline, and the world's observatories are equipped with instruments that reveal information about how objects and processes in the universe radiate, reflect, and absorb light across the spectrum. To get a complete 'picture' of an object, its origin and evolution, astronomers study it in as many wavelengths of light as they can, since each 'regime' of the EMS reveals something special about the object or process.

Not all observatories are *on* Earth. Some are in space, and others are actually at or on other planets. The biggest obstacle to getting a truly multi-wavelength view of the universe is our atmosphere – the blanket of air that sustains life on our planet. It absorbs some incoming radiation and also can blur the view if it's especially turbulent. Many professional-grade observatories on our planet are found on mountaintops or in deserts, and for the best view, they are outside the atmosphere. This last is true for facilities sensitive to gamma-rays, x-rays, and ultraviolet, as well as some aspects of infrared. They need to be well away from Earth's atmosphere to get the best science.

The Nature of Light

Since astronomy is the scientific study of light from objects and processes in the universe, it's reasonable to take a detour into what light is. People commonly refer to light as the visible light we see, which is understandable. It is, after all what humans and other life on Earth evolved to 'see'. But, we also detect other forms of light such as infrared and ultraviolet with our skin. Other life forms on Earth are sensitive to some things that humans cannot detect. Interestingly, life also gives off light by virtue of its temperature, in the form of infrared, or heat radiation. Some species exhibit 'bioluminescence', such as fireflies, jellyfish, tiny ocean organisms, and some deep-sea animals. Whether receiving, sensing, or emitting it, light is an important part of the universe, in all its many wavelengths and frequencies. And without light, we'd know very little about our own planet, let alone the rest of the universe. Observatories help us detect and study all that light.

What is light, exactly? In physics, scientists refer to each 'bit' of light as a photon or a quanta. Light can behave as a particle or a wave, but not both simultaneously. For example, light leaves the Sun as a stream of photons but it can also propagate and interact with matter as a wave. This admittedly confusing and seemingly contradictory double nature of light is sometimes described as the 'wave-particle duality'. Light has several properties of interest to astronomers: intensity, its direction of motion (its propagation), its frequency or wavelength, its speed (either through a vacuum or some other medium such as air or water), and its polarisation (direction of oscillation of a light wave).

Light is usually measured in terms of wavelength in most astronomy disciplines. Visible light (which we see with our eyes) ranges from 380 nanometres, at the violet end, to 740 nanometres at the red end. Ultraviolet light stretches from 10 to around 400 nanometres. However, in radio astronomy, the light is often measured in terms of frequency. So, for example, radio astronomers refer to MHz, KHz, and GHz – which refers to megahertz, kilohertz, and gigahertz, respectively. It may sound complex, however, remember that the unit 'Hertz' (which comes from the name of Heinrich Hertz, who in 1888 built the first radio receiver) describes one cycle in a sound wave. One cycle is 1 Hz. A million cycles is 1 MHz, a thousand cycles is 1 KHz, and a billion cycles is 1 GHz. The entire radio frequency 'band' in the spectrum includes frequencies between 30 Hz to 300 GHz.

In x-ray and gamma-ray astronomy the units keV, MeV, TeV, and GeV (for kilo-electron-volt, Mega-electron-volt, and so on) are used to express the energy of radiation streaming from such objects as galactic jets or gamma-ray bursters, or the energy of a neutrino or cosmic ray.

Combining Light Signals

The technique of interferometry is important, particularly in optical wavelengths as well as radio frequencies. It was first used in radio observations of the Sun in the mid-1940s. Essentially, interferometry combines the light (or radio frequencies) from two or more telescopes. The distance between the telescopes is called the baseline. That's also the diameter of the virtual telescope created by ganging together a number of telescopes. The more telescopes astronomers add, the more information they get about an object. In the case of the Event Horizon Telescope mentioned above, ten observatories were plugged together to do observations and get more detail about the event horizon of a distant black hole.

Radio Telescopes

Radio telescopes are the workhorses of non-visible astronomy. Such an instrument can be a large parabolic antenna, often referred to as a 'dish'. It gathers the incoming radio waves from cosmic objects and focuses them down to a collection of detectors and other instruments, just as mirrors reflect and focus visible light. Radio telescopes come in a range of sizes and designs. For example, the largest single radio dish in the world is the Five hundred meter Aperture Spherical Telescope (FAST) in China. The next largest is the Arecibo radio observatory in Puerto Rico, at 305 metres across. These two are built into natural valleys on Earth.

Objects in space give off radiation from all parts of the EMS. The ones that give off radio emissions are usually energetic, and the science of radio astronomy is the study of those objects and their activities. Radio astronomy reveals an unseen part of the universe we cannot detect with our eyes. This branch of astronomy began when the first radio telescope was built in the late 1920s by Bell Labs physicist Karl G. Jansky. He'd been assigned to work on a radio antenna sensitive to radio waves at a frequency of 20.5 MHz. He built a turntable to hold it, and then spent several months using it to detect signals, which he then sorted into noise from thunderstorms and some kind of unknown static hiss whose origin he couldn't pinpoint.

Jansky noticed the hiss was periodic – that is, it would rise and fall once a day. At first, he thought it might be radio noise from our star. But, the timing of the rise and fall didn't exactly correlate with the Sun. Instead, it seemed to follow a fixed point in the sky. So, he got out the astronomy star charts and determined the mysterious signal was coming from the heart of the Milky Way. Jansky published a paper about his finding in 1933, called 'Electrical disturbances apparently of extraterrestrial origin'.

Although this was a very significant finding, the work found little support or interest from the astronomy community – at first. Some years later, engineer Grote Reber built a radio telescope and confirmed what Jansky had found. Physicist John D. Kraus followed up with the first radio observatory, called the 'Big Ear', at Ohio State University, where the first official radio survey of the sky was conducted. Since Jansky's time, radio telescopes have popped up around the world, and there are even some proposals to put them on the far side of the Moon for the ultimate in 'radio quiet' observing. His work is always pointed to as the starting point of the field, and radio astronomers honoured his contributions by naming the standard unit for the strength of a radio signal as the 'jansky'.

These days, radio astronomy is expanding to radio-quiet places around the globe. Arrays can be found in the US, Europe, South America, and are under construction in Australia and South Africa. Single dish observatories continue to be used as well. The most famous one in England is the Lovell Telescope at Jodrell Bank in Cheshire, the third-largest single steerable dish in the world. It was brought online in 1957 and has been upgraded over time and is now part of the European Very Large Baseline Array, as well as the Multi-Element Radio Linked Interferometer Network (MERLIN) array of radio telescopes in England.

In West Virginia, the Robert C. Byrd Green Bank Telescope (called the GBT for short), is the largest fully steerable antenna in the world. It complements the ALMA and VLA arrays, and also works with other interferometers to get high-resolution radio views of the sky. New generations of astronomy students can train using this telescope, which replaced an older single dish that collapsed. It is also used in a project called 'Breakthrough Listen', which is scanning the skies for signals which could be from distant civilizations. The GBT is located in a radio-quiet area where no one is allowed to use cell phones and other electronics, and even the vehicles are free of electronic components. This reduces the radio frequency interference that would otherwise plague the telescope.

The Five hundred meter Aperture Spherical Telescope (FAST), mentioned above, was first used in 2016 and has already made discoveries during its testing and commissioning. It's not the only single-dish facility set into a natural geographic feature. Most people are familiar with the venerable Arecibo Observatory in Puerto Rico, which was until the completion of the FAST instrument the world's largest single-aperture radio telescope. Set in the hills of Puerto Rico, this 305-metre-wide telescope takes advantage of a natural depression in the ground to do radio astronomy, atmospheric science, and radar astronomy. Most people know of this famous facility from the movie *Contact* (which also featured the VLA). In September 2017, the dish sustained damage from Hurricane Maria (which devastated Puerto Rico and nearby islands). It has been repaired and remains in use today. Most of these facilities exist well away from the radio frequency pollution of modern life, on deserts and mountains where radio astronomy can flourish.

Many people have seen or have visited the Lovell Telescope at Jodrell Bank in England, or arrays, such as the Very Large Array dishes in New Mexico in the US. Those huge dishes can detect whisper-quiet signals from across space. Other installations look more like a collection of metal rods and wires attached to each other and spread out across a landscape. That's the case for the Murchison Wide Field Array (MWA) in Western Australia. It is designed to capture very low frequency signals from objects in space. Its sister installation, called the Low-frequency Array (LOFAR), located in Europe does the same thing. Another one, called the Square Kilometre Array (SKA), is still in the planning stages, and will create a virtual kilometre-sized radio telescope 'farm' of dishes. We will examine some of these in greater detail in later chapters.

Arrays are providing some of the most exciting work being done in radio astronomy these days. In addition to the Very Large Array, Murchison Widefield Array and the Low-frequency Array, the Atacama Large Millimeter Array (ALMA), high in a Chilean desert region, and the Event Horizon Telescope, chain together many telescopes tuned to the same frequency ranges. For the more widespread arrays, the technique of very large baseline interferometry (called VLBI for short) employs radio telescopes spaced out across continents to simultaneously observe an object. The data from those detections are correlated and combined into larger data sets for analysis, and can be used to create an 'image'.

VLBI observations simulate the ability to see detail in a celestial object that would otherwise require a telescope the size equal to the separation between the elements in the array. For example, the Very Large Baseline Array in the United States produces a 'virtual telescope' more than 8,000 kilometres in extent. The technique allows it to look at very small areas of space to pinpoint variations in signals from strong radio sources, for example, or to zero in on the region around a supermassive black hole.

X-ray Telescopes

An x-ray telescope doesn't look much like a typical telescope at all, even though it uses mirrors. That's because it uses them in a different configuration. X-rays are highly energetic radiation and are guided to ricochet off highly polished mirrors embedded inside each other, like nested barrels. Chandra X-Ray Telescope's mirrors are a good example of the design. They channel the x-rays toward the detectors and other instruments that measure the strength of the x-rays and other characteristics. Because x-rays are absorbed or scattered by Earth's atmosphere, x-ray astronomy must be done out in space. Along with Chandra, a number of other x-ray detecting telescopes have orbited Earth, including XMM-Newton, NuSTAR and others. In the early days of x-ray astronomy, scientists used rockets and balloons to loft their detectors high enough to get good measurements. Good examples of x-ray-emitting objects are the Sun, the Crab Nebula, and active galactic nuclei.

Gamma-ray Telescopes

Gamma-ray telescopes contain specialized detectors and spectrometers. Gamma rays stream from such nearby events as solar flares as well as from distant pulsars, supernovae, and other energetic events. Astronomers construct detectors that record the 'ionization' taking place when gamma rays interact with matter. When the gamma-ray collides with an electron, it transfers all or part of its energy; the ionized electrons then collide with other atoms and that liberates more electrons. Depending on the type of instrument doing the capturing, the liberated electrons are counted, or measured indirectly by a scintillation detector. The data about the gamma-ray and its interactions is then turned into an electrical pulse with a voltage proportional to the energy deposited in the medium used to collect it. To put it more simply, the gamma-ray interacts with electrons and the resulting 'outbursts' are measured and turned into data. That data is then analysed for what it can tell astronomers about the object emitting this radiation.

Telescopes in Space

Space-based observatories also contain a variety of telescopes sensitive to different 'regimes' of the electromagnetic spectrum. Each has instruments designed for the wavelengths of light scientists are most interested in studying. If required, the spacecraft carries along coolants for infrared and other detectors.

Hubble Space Telescope, for example, has used visible-light detectors, as well as ultraviolet- and infrared-sensitive instruments. The Spitzer Space Telescope was outfitted with all infrared-sensitive instruments. The Chandra X-Ray Observatory is, as its name suggests, a strictly x-ray-sensitive machine. It is joined by others, among them Exosat and NuSTAR. These, and other sister satellites launched by agencies around the world, operate in space largely because the wavelengths they are built to study are absorbed by Earth's atmosphere. Simply put, they can only do their best work well above our blanket of air.

The days are long past when astronomers simply looked through the eyepiece and drew what they observed or made lengthy descriptions in notebooks. Today, professional observatories have an array of instruments and hardware to record and study light from distant objects. Many amateur observatories (some of which are quite sophisticated) also use some of the latest technology that was once only available to the pros.

Building Observatories

Over the centuries since the telescope was invented, astronomers have faced great obstacles to get their instruments in place. Some of these trials and tribulations are mentioned in other chapters, but in the main, observatory builders have dealt with bad weather, bad roads, broken equipment, cracked mirrors, wars, famines, fires, and other tragedies. Nearly every facility has faced some major challenge. The builders of the Mount Wilson Observatory, for example, faced horrific storms, tourists, and lousy roads simply to establish a viewing space on top of a mountain. Once they did, they also faced a tremendous challenge getting their mirror created, ground, and transported up the mountain using mule- and horse-drawn wagons.

Altitude was (and remains) a major obstacle for the builders of telescopes on Mauna Kea, in Chile, and on the Atacama Desert. During the construction of the Gemini telescopes, workers were strictly limited in the number of hours they could operate in the forbidding environment where their observatories were being built. Astronomers who go to observe in these places are usually asked to arrive a few days early to acclimate to conditions before operating telescopes. In some very high-altitude observatories, oxygen supplies and specialised 'Gamow bags', which serve as portable hyperbaric chambers, help those struggling with altitude sickness.

Fires have threatened many a remote observatory. Periodic forest fires in southern California have sent astronomers and technicians fleeing from Mount Wilson and Palomar Mountain. In Australia, a horrific brush fire swept over the historic Mount Stromlo Observatory, destroying most buildings and all but one telescope.

Wars and invasions have disrupted observatory operations as well. During the Second World War, the Pulkovo Observatory near St Petersburg in Russia came under sustained attack by the Germans. Astronomers managed to save parts of the telescopes, but the buildings were destroyed and the library and archives were largely burnt. The invasion of China by Japan prior to the Second World War affected the Purple Mountain Observatory, which shut down and transferred some instruments and people to Kunming until 1946. News reports and first-hand anecdotal information from Kuwait detailed the destruction of the Kuwait National Museum with its observatory and planetarium just prior to the United States entering the region during the Gulf War. Further back in history, the Great Library of Alexandria contained laboratories and an observatory, as well as the collection of written works and botanical gardens, and a zoo. According to legend, this complex was destroyed in 48 BCE by Julius Caesar, although this is still a subject of debate among historians.

The biggest threat to observatories in modern times is light pollution and radio frequency interference. The enhancements of modern civilization – lights, cellular communications, radio and TV transmissions – all put astronomy at risk. To combat this, as we'll see in later chapters, astronomers simply moved their observatories to dark sky sites. They still face challenges from constellations of satellites crossing the skies all night, but those, at least, can be worked around. As long as there are ways to observe, astronomers will find a way to pursue their science using observatories.

Dark Skies Advocacy

To combat the rise of light pollution, astronomers formed the International Dark-sky Association, an advocacy group educating the public about the overuse of lights at night. Its members around the world teach about the challenges to human health and wildlife that excess light pollution poses. In addition, they campaign for dark-sky areas for both amateur and professional astronomers. One of the group's main outreach tools is the designation of specific places as 'dark sky preserves'.

Within these areas, people are encouraged to use lighting wisely (if at all), and to help preserve the sky access of current and future generations. As mentioned, light pollution is the number one reason why observatories are located well away from cities, and why some observatories have had to move from their original sites. The establishment of dark sky preserves in places such as Flagstaff, Arizona, and others, recognises the need for access to the sky, whether one is a professional astronomer or a backyard observer.

The Instruments Inside Observatories

Beyond telescopes, modern astronomy makes use of additional equipment to record and analyse light from space. There are sets (or suites) of instruments that attach to the telescopes to help with imaging as well as dissecting light into its component wavelengths for deeper study.

Regardless of the type of light an observatory is built to gather, the astronomers need ways to detect the light, convert it into data, and deliver the information to computers. They also require equipment to record images and data. Observatories are equipped with detectors specific to their wavelength ranges.

Recording the sky through the observatory eyepiece has always been a big part of the astronomical observatory's contribution to science. Before the invention of the camera and photography, the only method that astronomers had to record what they saw was to draw it from memory. In prehistoric times, rock art was the method. Very early astronomers created paintings, carved their data on clay tablets, or sketched what they saw on the walls of their caves. Galileo drew on parchment, which is how we know about his observations of the Sun, Jupiter, its moons, and other objects.

Today, the camera of choice for optical and near-infrared work is digital. It receives the incoming light and turns it into electrical impulses and eventually into data. The camera may use a charge-coupled device (CCD) or a complementary metal oxide semiconductor (CMOS) chip to convert the incoming light. Both are light-sensitive integrated circuits that employ picture elements, or pixels, to gather light. Think of them as tiny light buckets. When photons of light travel through the telescope they fall onto the pixels. In a CCD, a converter turns each pixel's value into a digital number. They do this by measuring the amount of charge at each pixel and converting it to binary numbers. In the CMOS, transistors are used to amplify the charge at each pixel and move it through wires to other parts of the camera. Modern consumer cameras, including those in our smartphones and 'point and shoot' instruments contain these arrays. They work the same way a telescope's detector works.

Before the advent of CCDs, observatories used film cameras and photographic plates to record distant sky objects. The introduction of cameras into the observatory was a huge step forward for astronomy, and provided time-exposure images of distant objects the eye could never see. Digital cameras aren't limited to professional observatories; they are a staple for serious amateur astronomers and astrophotographers.

Photometers are devices used to measure the intensity of light streaming from a target. Light comes into the telescope and then is passed through to the photometer, which usually has a CCD chip that converts the light into an electric current. Among other things, photometers are used to measure brightness changes in such objects as variable stars, which slowly change their luminosity over time. Photometric studies can give information about an object's structure, age, distance, and temperature.

Spectral Instruments
Astrophysics (the study of the physics of stars and galaxies) depends on the study of light in great detail. Imaging is certainly one important way to do this. To get at the heart of what makes a distant object 'tick', it's important to dissect its radiation in a way similar to sending light through a prism. This technique is called spectroscopy, and requires instruments that can separate the incoming wavelengths and frequencies into their tiniest components. Buried inside the data is information about the object's chemical makeup, speed through space, magnetism, temperature, and other important characteristics. The most commonly used instruments to do this work are called spectroscopes and spectrometers. They come in various types and are attached to the telescope after the prime focus. Electromagnetic radiation is guided in and the instruments then split it into its component wavelengths.

The Hubble Space Telescope had as part of its original instrument array the Goddard High Resolution Spectroscope (GHRS). It was sensitive to ultraviolet wavelengths and used specialised gratings to spread out the UV light. From there, astronomers could analyse the spectra to understand an object's motion, chemical makeup, and other characteristics.

At the Gemini Observatory, for example, where astronomers routinely do both optical and infrared observations, spectroscopy is done with the Gemini Multi-object Spectrograph (GMOS). This specialised instrument allows astronomers to make detailed studies of light from celestial objects such as a supernova in a distant galaxy, or measure the redshifts (the velocity at which they appear to be travelling away from us) of many galaxies in a field of view. Spectrographs are a commonly used instruments in almost all observatories today, both ground-based and in space.

Adaptive Optics

Earth-based observatories, even those high atop mountains around the world, have to contend with a problem called 'atmospheric aberration'. It's the effect that motions of air in our atmosphere have on the light streaming through from distant objects. This aberration blurs the images and limits the ability to see objects in sharp detail. For centuries, there wasn't much anyone could do about it, except hope for good, stable weather. Then, starting in the late 20th century, a technology called 'adaptive optics' found its way from military use to civilian application in observatories. Essentially, an adaptive optics system helps remove the effects of atmospheric aberration. The results give sharper, clearer images and spectra of the sky.

Adaptive optics systems use deformable mirrors that are controlled by computers. They correct for distortions caused by our atmosphere and are mounted alongside the main telescope in an observatory. The systems require a fairly bright reference star. If no natural star is nearby in the field of view, then the system can be programmed to shoot a laser beam to a point in the upper atmosphere. That creates a reference point of light, a 'laser guide star' bright enough for the adaptive optics sensors to see. The system focuses on the guide star and studies (in real time) how its appearance changes due to motions in the atmosphere. The data from the guide star observations is then used to 'subtract' or cancel out the motions and aberration from science observations. The application of adaptive optics results in much-improved images and data from the ground-based systems. In some cases, the images are very close to (or sometimes better than) the quality of those taken from space.

Using an Observatory

Modern observatories allocate observing time based on a variety of factors. Each observatory has its own rules and requirements, and generally speaking, those factors are taken into account as proposals come in for each facility. For most observatories, observing opportunities are announced at least once a year, sometimes twice. The deadlines are communicated to the observing communities they serve, and if there are any specific guidelines, those are also spelled out.

Each observatory has a constituency it serves and members of its research institutions attached to the observatory are allowed to apply for time. They are generally supported by research grants which cover their salaries (and that of other team members, some of whom may be students) and other expenses related to the work they're doing. This doesn't mean outsiders aren't ever allowed to apply to use telescopes their institutions may not be part of. In some cases, they are. Often enough, large observing teams include members who may not be part of the institutions attached to specific observatories. Sometimes, those teams also include very talented amateurs participating in observing programmes.

The proposed targets for observation have to be 'do-able' using the specific observatory's equipment. The proposing scientist (called the principal investigator, or PI) needs to be familiar with the facilities and what the instruments can do. It's particularly important for the astronomers to know, for example, what wavelength range they want to observe their targets in. A proposal to observe changes in the clouds of Uranus, for example, may specify optical observations and perhaps some infrared spectroscopy to get some idea of the chemical makeup of the clouds or take temperature measurements. The observers also have to give a length of time for the observations: do they want several hours each night for several nights? One session of observing on one night? Each specific target will have optimal times for observations.

Once the PI has decided on the particulars of their observing proposal, they fill out a form (usually online) with the details and the science justification for the request. That form goes to a time allocation committee (TAC), which evaluates the proposal. As part of deliberations, committee members discuss the scientific merits, the 'load' it will put on the observatory's facilities, and other factors. Then, they decide whether to accept the proposal and assign a time on whatever telescope and instruments have been requested.

As an example of how it works for one facility, the author and her advisor proposed a series of imaging observations for Comet C/1995 O1 (Hale-Bopp) for late 1996. For the project, the team needed to look at the comet for several nights running, using the University of Hawaii 2.3-metre (88-inch) telescope on Mauna Kea. The proposal specified the length of each set of observations (each several minutes long). The proposal was submitted in the spring of 1996, and the TAC accepted it for observations. Some members of the team were already at the University of Hawaii and part of its constituency. It was judged acceptable to let the author's proposal join in with another one the Hawaii team had submitted. The author's team had requested time on the telescope in November of the same year, at a time when the comet's position was favourable for observations.

Those comet observations were not the only work done on those nights. The Hawaii team members were using the telescope the same nights to do light-curve studies of asteroids. The scheduled observations complemented each other, and allowed a very efficient use of the telescope each night. The images were stored on a computer at the University of Hawaii, and also copied to a hard drive for the author to take back to the university for further study.

Receiving Data and Images
Until relatively recently, astronomers nearly always went 'up the mountain' to do their observations. That is changing with the advent of queue observing, as well as robotic observing. Queue mode is a way of scheduling the telescopes at a given observatory. It's where observing runs are carried out by the telescope operators (who were formerly called 'night assistants') without requiring the proposing astronomer to be present. It's a more efficient way to schedule telescope time and is in use at a number of facilities and allows for flexibility 'on the mountain' to account for changes in weather, observer requests, etc.

Of course, the PI may wish to travel up to the mountain – perhaps the observations require their presence, for example. Most observatories have visiting astronomers' quarters where they can get a place to sleep, access to computers, and three meals a day. The quarters at Mauna Kea, on the Big Island of Hawaii are called the Onizuka Center for International Astronomy (*Hale Pohaku* in the Hawaiian language). There, astronomers can spend a few days acclimating to the 2,800-metre (9,300-foot) altitude before heading up to the summit, which is 4,205 metres (13,803 feet) above sea level. A similar arrangement exists for astronomers arriving to use telescopes at high altitudes in Chile and other facilities. The European Southern Observatory, for example, has a guest house called the *Paranal Residencia.*

When the time comes for the observations, the night assistant (and the PI, if she or he wants or needs to be present) executes the observing programme on the requested instruments. The resulting data are stored (usually with some minimal processing) where the observation team can access the information. The PI then collects the data and distributes it among his or her team members for further analysis.

The time allocation process for observatories is roughly the same for both ground-based and space-based facilities. For space-based observatories, however, the observers don't have to travel anywhere for the actual observations. Their data is collected by the specific space platform and sent back to Earth, where it is processed lightly and then put into storage for the principal investigator to retrieve.

Space observatories such as Hubble Space Telescope are in very high demand, and there is only so much time available. Generally speaking, the HST time allocation committee receives four to six times the numbers of proposals than the telescope can handle and its available observing time is highly oversubscribed. In practical terms, that means the proposals need to be for targets and investigations which can benefit only from time on HST. For example, when the New Horizons team was searching the Kuiper Belt (the outer solar system) for a second target for its spacecraft, it applied for time on HST because that observatory had the best chance of finding something along the spacecraft's trajectory after its flyby of Pluto.

Some observing proposals for both ground- and space-based observations are made for what's called 'target of opportunity' (TOO) time. This covers objects such as newly discovered comets or moons or a supernova explosion. In 1987 a supernova exploded in the Large Magellanic Cloud in the southern hemisphere skies. Within a day, TOO proposals for SN 1978a were being executed on various telescopes around the world. Once the Hubble Space Telescope was launched, it, too, was aimed at the fading supernova remnant (only three years later). Another good TOO example was the discovery of Comet Shoemaker-Levy 9 (D/1993 F2). It was headed straight for

Jupiter and astronomers mobilised to observe it from observatories around the world, as well as with HST.

With the advent of sky surveys, astronomers are being alerted to a number of other events, such as supernova explosions and gamma-ray emitters. Once one of these extremely energetic outbursts is spotted, observatories can swing into action to make optical observations as the gamma-ray burst (GRB) quickly fades. This same procedure occurs when a nova is spotted, or any other short-lived energetic event.

The world's professional observatories are popular places and it's safe to say that most of them get many more proposals than they can handle in a given 'semester' of observational time. It falls to their time allocation committees to discern the best use of these limited resources. Each observatory is fully booked year in and year out, and aside from weather considerations or equipment failures, they continually 'crank out' good science.

What Do Observatories Study?

Astronomy and astrophysics are the sciences concerned mainly with the study of the solar system, stars, extrasolar planets, nebulae, galaxies, and the origin and evolution of the universe. Those sciences break down into sub-disciplines such as planetary science, solar physics, and cosmology. Over the past centuries, astronomers have learned that planets exist, they have moons and rings, that asteroids and comets date back to the earliest epochs of solar system history. They've found planets around other stars, and learned nearly all stars host planetary systems, some Earth-like. Astronomers have found the birth places of stars and the remains of dead stars, not only in our galaxy but in others. And in the large-scale universe, observatories have found structure in the distribution of galaxies, evidence of dark matter, a mysterious 'something' called dark energy, and looked back to the earliest times in cosmic history. It's a heady science which has shown us our place in an expanding universe.

Astronomers can opt to observe single objects or groups of objects, such as a portion of a nebula or star cluster, comet, planet, or galaxy. Or, they can band together and do large-scale surveys of the sky in all the different wavelengths of light. For example, The Sloan Digital Sky

Survey has studied a large area of the sky, gathering optical data and doing spectroscopy on objects using a dedicated telescope. The Great Observatories Origins Deep Survey (GOODS) combined observations from Hubble Space Telescope, the Spitzer Space Telescope, and the Chandra X-Ray Observatory as well as large ground-based telescopes to look at distant galaxies and study their formation and evolution. Redshift surveys, which study selected portions of the night sky to measure the redshifts (that is, the change in wavelength of light from an object as it moves through space) of distant galaxies, clusters of galaxies, and quasars. The information from these surveys is used to determine their distances, and in many cases, allows astronomers to create a three-dimensional map of the objects they surveyed. In addition, more recent redshift surveys of distant quasars and galaxies are being used to map the extent and distribution of dark matter and the contribution of dark energy to the expansion of the universe.

Astronomy is, as noted elsewhere, one of the oldest of the sciences humans do. As we'll see in Chapter 2, early civilizations were involved in studies of the stars, ranging from Asia and the Middle East to cultures as far-flung as the earliest Native Americans and Polynesian navigators. However, their interest was not purely scientific. It was more connected to timekeeping, calendar-making, and social and religious rituals.

In modern times, astronomy and astrophysics concentrate on both observational discoveries as well as theoretical models to help scientists understand the universe. For the purposes of this book, we concentrate on the observational science advanced by observatories equipped with state-of-the-art telescopes and detectors. To understand why astronomers use so many different types of telescopes, it's worthwhile to look more closely at the objects they study. Some objects require more than simply looking at them in optical light.

Brightness in Astronomy

One of the most important aspects of understanding a star or other object in the universe is knowing its brightness. How bright is it? How bright is a planet or a galaxy? Astronomers express celestial brightnesses using the term 'luminosity'. It describes the brightness of an object in space.

Stars and galaxies also give off various forms of light, as we've seen earlier in this chapter. What *kind* of light they emit or radiate tells how energetic they are. If the object is a planet, it doesn't emit light; it reflects it. However, astronomers also use the term 'luminosity' to discuss planetary brightnesses.

Stars radiate in very broad sets of wavelengths, from the visible to infrared and ultraviolet; some very energetic stars are also bright in radio and x-rays. The central black holes of galaxies lie in regions that give off tremendous amounts of x-rays, gamma-rays, and radio frequencies, but may look fairly dim in visible light. The heated clouds of gas and dust where stars are born can be very bright in the infrared and visible light. The new-borns themselves are quite bright in the ultraviolet and visible light.

The greater the luminosity of an object, the brighter it appears. It also turns out that an object can be very luminous in many wavelengths of light. For example, a bright young star can be seen in visible light, but it may also give off extreme ultraviolet light, or x-rays. A bright galaxy core can also be very luminous in radio frequencies, while its outer arms might be very bright in infrared light. In these cases, luminosity also tells astronomers how much energy is being emitted by an object in all the forms of light it radiates (visual, infrared, x-ray, etc.). Astronomers study the different wavelengths of light from celestial objects by taking the incoming light and using a spectrometer or spectroscope to 'break' the light into its component wavelengths. This method is called 'spectroscopy' and it gives great insight into the processes which make objects shine.

We can get a very general idea of an object's luminosity simply by looking at it. If it appears bright, it has a higher luminosity than if it's dim. It's pretty logical. However, that can be deceptive. A distant but very energetic star can appear dimmer to us than a lower-energy but closer one.

Astronomers figure out a star's luminosity by looking at its size and its effective temperature. The effective temperature is expressed in degrees Kelvin, so the Sun, for example, is 5777 Kelvins. A distant quasar could be as much as 10 trillion degrees Kelvin. Each of their effective temperatures results in a different brightness for the object. The quasar, however, is very far away, and so appears dim.

The luminosity which matters when it comes to understanding what's actually happening inside an object, from stars to quasars, is its intrinsic luminosity. That's a measure of the amount of energy it actually emits, regardless of where it lies in the universe. Knowing an object's intrinsic luminosity is a way of understanding the processes inside the object which make it bright.

Luminosity and Magnitude

Another way to understand and measure an object's brightness is through its magnitude. Observers often refer to an object's brightness in comparison to a standard magnitude scale. The magnitude takes into account an object's luminosity and its distance. A second-magnitude object is about two-and-a-half times brighter than a third-magnitude one, and two-and-a-half times dimmer than a first-magnitude object. The lower the number, the brighter the magnitude. The Sun, for example, is magnitude -26.7. The star Sirius is magnitude -1.46. It's actually 70 times more luminous than the Sun, but it lies 8.6 light-years away and is slightly dimmed by distance.

Apparent magnitude refers to the brightness of an object as it appears in the sky as we observe it, regardless of how far away it is. The absolute magnitude is really a measure of the *intrinsic* brightness of an object. Absolute magnitude doesn't depend on distance; the star or galaxy will still emit a certain amount of energy no matter how far away the observer is. That makes it more useful to help understand how bright and hot and large an object really is.

Stars

When one goes outside on a clear, dark night, the first thing that catches the eye is the starry sky. There are about 9,000 stars visible to the naked eye across the entire sky, and many more if one gathers more light and magnifies the view with a telescope. Early observers had many ideas about what stars are. Some thought they could be campfires or the spirits of the dead, or in a loftier way, some kind of deities. No matter what people thought of them, the stars and planets were part of an unreachable realm. They still are, but the difference between those early perceptions and what we know about the universe today is the result of centuries of work by astronomers applying scientific principles to understanding what's 'out there'.

So how did they learn about stars? The closest one is the Sun and provides a fine example to study. It's a typical middle-aged yellow dwarf star, one of many in the galaxy. It, and all the other stars in the universe, shine by the process of nuclear fusion. That is, deep inside a star, in the core region where temperatures and pressures are incredibly high, atoms of hydrogen are fused together to form helium. The process emits heat and light. No matter how old a star is, it does this at some point in its life. As the Sun ages, it will eventually run out of hydrogen and will start fusing helium in its core to make carbon and oxygen. For stars far more massive than the Sun, 'fuel loss' can happen a number of times as the star ages.

The lives of stars comprise one of the main topics in astrophysics – the application of the laws of physics to astronomical objects such as stars, nebulae, and galaxies. To advance our knowledge of stars, astronomers and astrophysicists study starlight very carefully. In essence, they use specialised instruments such as spectrographs and spectrometers to 'take the light apart'. Each wavelength of light supplies crucial data about the star, its characteristics, motions, and evolution.

Planets

Most skygazers are familiar with the appearance of the planets. These worlds in our own solar system originally puzzled early observers. That's because they seemed to move across the backdrop of the 'fixed' stars. The term 'planets' comes from the Greek word *planetes*, which means 'wanderers'. It wasn't until the advent of the telescope, beginning with Galileo Galilei's first observations of Jupiter and its moons, that observers really began to comprehend these objects are worlds in their own right.

The brightest planets – Mercury, Venus, Mars, Jupiter, and Saturn – are easily visible to the naked eye. Uranus, Neptune, and distant Pluto (technically a dwarf planet), can only be seen with telescopes. Even good backyard-type telescopes can reveal some features such as Jupiter's Great Red Spot, Mars's polar caps, and Saturn's rings. However, it's important to remember that even the strongest telescopes cannot resolve every feature on a planet. For very 'up-close and personal' looks at a planet, researchers must send spacecraft to the planets. Scores of robotic explorers have swarmed around the planets. They supply everything from high-resolution looks at surface rocks to detailed studies of the gases in the planets with appreciable atmospheres.

Earth-bound observatories continue to study planets; notable examples range from the Keck observations of Uranus to Hubble Space Telescope's observations of Mars and Jupiter. In addition, telescopes at institutions such as the European Southern Observatory, Gemini, and others, regularly check out events at multiple wavelengths. They study auroral displays and impacts on Jupiter, and changes in the clouds of Saturn and Neptune.

Planets aren't the only things observatories search out in our solar system. They also routinely study comets, asteroids, moons, and dwarf planets. Every world that can be observed has, at one time or another, had some kind of detector aimed at it to record its characteristics and activities at multiple wavelengths.

Extrasolar Planets

In 1988 astronomers began detecting extrasolar planets. These are worlds orbiting other stars, and are often dubbed 'exoplanets'. Today, thousands are known and more are waiting to be confirmed. Not only have ground-based observatories been used for this work, but Hubble Space Telescope has found its share of them too. The most extensive survey was done by the orbiting Kepler space telescope. It aimed only at a small part of our sky, and found candidates around many stars in its field of view. Joining it in space were the European Space Agency's COROT satellite as well as the currently operating Gaia, ASTERIA, and TESS satellites. Today, exoplanet searches and confirmations are being done by these space-based telescopes as well as a number of ground-based telescopes. It's an 'all hands to the pump' effort to both discover and confirm the ongoing rush of exoplanet discoveries.

The study of exoplanets is not easy. These worlds are largely hidden in the glow of their parent star's light. That makes them virtually impossible to observe from the ground without special instrumentation. Space-based surveys don't have to contend with Earth's atmosphere and can be programmed to block out (occult) the star so that planets can be detected.

Astronomers use several techniques to find them in the first place. This includes measuring the 'wobble' of a star in its orbit which might be due to the smaller gravitational influence of its planets. Astronomers have also perfected a technique that lets a telescope 'watch' via a sophisticated 'light meter' as a planet crosses between us and its star. This 'transit' method is what Kepler used to find so many exoplanet candidates.

Once planet candidates are found, astronomers follow up with ground-based observations to confirm they are actually planets. It's painstaking work, but the rewards are great. Today, we know of a variety of planets out there, ranging from Earth-size analogues to super-Earths, hot Jupiters, 'super-Jupiters', hot Neptunes, Goldilocks planets (those orbiting at the right distance from their stars to support liquid water on their surfaces), and many others. In the process of identifying these worlds, exoplanet experts are also improving theories about how planetary systems form and evolve. Extrapolating the number of planets in the entire Milky Way from the discoveries made so far shows that there could be as many as 100 billion planets just in our own galaxy alone. The Holy Grail of exoplanetary studies is to find an Earth-like planet capable of supporting life.

Nebulae

Our galaxy and many others are studded with clouds of gas and dust. The common term for such a region in space is 'nebula' and it is a Latin word for 'cloud'; the plural is 'nebulae'. Most of these clouds are made mainly of gases interspersed with dusty grains. The most famous one in our skies is the Orion Nebula, a region where stars are being born as we speak.

There are different types of nebulae and some regions can contain several types. Many can be called 'diffuse' nebulae because they are quite well-extended but usually don't have defined shapes. H II (for ionized hydrogen) regions typically include other gases as well as dust.

Some nebulae shine from the light of stars embedded inside. Those stars provide energy to the surrounding gases and cause them to glow. Such regions are called 'emission' nebulae. Others simply reflect the light of nearby stars and are called 'reflection' nebulae. There are also dark nebulae, which are very dense interstellar clouds that consist of cold gas and dust grains. They neither reflect nor emit, and they absorb the light of anything inside them or nearby.

There are also planetary nebulae, which result when sun-like stars puff off their outer atmospheres into space as part of the aging process. Supernova remnants are the leftovers of massive stars which blow up and spread their remains into space. These star-death nebulae are the ultimate 'recycling' mechanism for the elements created inside stars. Those elements join the hydrogen (and helium) created in the Big Bang, and enrich the universe. The insides of stars supply the materials needed to create new stars, planets, and life.

Observatory telescopes focus on these regions of space to supply images and data that help astronomers understand all aspects of star life, from birth to death. Starbirth nebulae are studded with hot young stars that pump out ultraviolet radiation in addition to their optical light. Their intense radiation ionizes (pumps energy into) surrounding clouds of gas and dust, and in many places, the same radiation also 'eats away' the birth nebula. Astronomers see their optical and infrared glows in the Orion Nebula, as well as the Eagle Nebula in the constellation Serpens, and in similar nebulae in the nearby Large and Small Magellanic Clouds (satellite galaxies to the Milky Way). There's no doubt they also exist in galaxies throughout the universe and each tells a story of stellar evolution.

Star death in the form of planetary nebulae and supernova explosions gives astronomers multi-wavelength insights into the rich chemical mix blown away during the star's aging and death processes. Currently, astronomers are watching such stars as Betelgeuse (alpha Orionis) and the Eta Carinae star pair as they age. Betelgeuse is an aging supergiant and Eta Carinae is a luminous blue variable which has been steadily brightening and dimming over the centuries. Both are expected to explode within the next few million years. Continued observations help document the lengthy aging process they are experiencing.

The aftermath of star death isn't just a chemical enrichment story. The really massive stars, the ones that die in supernova and hypernova explosions, leave behind neutron stars and black holes. These are tremendously massive objects and they affect space and time in ways astronomers really only began to understand in the last decades of the 20th century. A black hole is 'easy' enough to understand: a region of space packed so densely with material its gravitational pull sucks in everything that draws near. This includes light, as well as gas, dust, other stars, and planets. There are stellar-mass black holes as well as supermassive ones at the hearts of galaxies. Some of the most active supermassive black holes are the main power sources for quasars.

Neutron stars are created when a supermassive star explodes as a supernova. The core is left behind, and is made up solely of neutrons, packed down as tightly as physics allows. If a neutron star was much denser, it would become a black hole. Recently, the collisions of neutron stars were found to create gravitational waves, which are incredibly difficult to detect. The Laser Interferometer Gravitational-wave Observatory (LIGO), which we discuss in Chapter 6, was built to sense these gravitational ripples, and is the first wave of a new branch of astronomy called gravitational wave science.

Star clusters are also part of astronomy's purview. They can be 'open' clusters that are simply traveling through space together (such as the Pleiades and Hyades clusters), or they can be massive 'globe-shaped' concentrations of stars called globular clusters. The main interests in cluster studies include their motions, evolution, and ages. For example, the Pleiades and Hyades clusters are thought to have been born in the same stellar nursery and will eventually spread apart as they travel through the galaxy. Such open clusters are found throughout a galaxy. A globular cluster, on the other hand, orbits the core of a galaxy, and is largely made up of stars that are as old as, or perhaps even older than its galaxy. The Milky Way, for example, has at least 152 globulars and there are probably another thirty or so hidden by clouds of gas and dust. Most globular cluster stars are considered 'low metal'; that is, they are largely hydrogen and helium and have very little of the heavier elements (called 'metals' by astronomers) carbon, nitrogen, and oxygen. Most astronomical studies of these clusters focus on detecting individual stars at their hearts, monitoring stellar interactions in their crowded environments, the evolution of stars within clusters, and determining the metal content and ages of their stars. In addition, astronomers are working to find out if black holes can exist in the cores of clusters.

Galaxies

Galaxies are defined as systems of stars, gravitationally bound together along with nebulae, interstellar gas and dust, as well as black holes, and dark matter. They are very different from star clusters, particularly those just discussed. They are generally smaller and consist almost exclusively of aging stars.

Our own Milky Way, along with its composite stars, nebulae, and black hole at its heart, has provided astronomers with a wealth of information about what to expect in other similar type galaxies. The Milky Way is a barred spiral galaxy, one of many in the universe. There are other types of galaxies, such as lenticulars, ellipticals, and irregulars, as well as dwarf spheroidals. One serious area of study seeks to understand the evolution of galaxies by observing as many different types as possible. It seems that galaxies develop hierarchically; that is, they start small and grow larger through cannibalisation of and mergers with other galaxies. This is why observations of the earliest galaxies is important, so astronomers can see them in their infancy and trace their growth throughout cosmic time. For such studies, infrared observations are crucial.

The Milky Way itself grew by mergers and interactions, and is still cannibalising smaller galaxies in modern times. The discovery of the Milky Way's voracious gobbling of smaller galaxies is an ongoing one. The observed and deduced history of own galaxy is helping astronomers make sense of other galaxy mergers they see in the distant universe. The reverse is also true: what they see 'out there' helps them understand how our own galaxy continues to evolve and grow by mergers.

Beyond the Milky Way, astronomers study the actions of galaxies (such as their rotation rates) and categorise them by shape, size, cluster membership, activity in their central regions, and whether or not they are interacting with other galaxies. Observatories are needed to take high-resolution images of nearby galaxies, and astronomers also study them with a large array of other instruments help dissect the light streaming from the individual galaxies.

Some major discoveries in astronomy stem from the study of galaxies. Among them is the discovery that galaxies are actually not part of the Milky Way. This was the prevailing view before Edwin

P. Hubble started to look at a variable star in what he and others called the 'Andromeda Nebula'. They referred to such objects as 'spiral nebulae' and posited they were simply part of the Milky Way. Hubble's work on the Cepheid variable Vɪ in Andromeda in 1923 proved it was so far away that the nebula containing it was actually a distant galaxy in its own right. That single observation changed our perception of the universe, almost overnight. Today, astronomers know the universe contains trillions of galaxies, and through studies made by observatories, also know galaxies were some of the first large-scale objects to form after the Big Bang.

Astronomers historically observed galaxies in visible light since that's all their telescopes could detect. However, with 20th-century advances in multi-wavelength astronomy, they have been getting a more accurate idea of activity within galaxies. This includes using infrared instruments to study star birth regions and the distribution of dust within a galaxy. They also use ultraviolet, x-ray and radio instruments to them track more energetic activity in galaxies. That includes supernova explosions, star activity, and the action of material swirling around central supermassive black holes. A good example comes from radio astronomy, which in 2019, used a world-wide array of radio telescopes to study the black hole at the heart of the galaxy M87. The first-ever image of a black hole was a huge breakthrough and showed the power of telescope arrays and associated computational abilities.

Clusters and Superclusters of Galaxies

Galaxies were once thought to travel the universe alone. We now know that most (if not all) galaxies in the observable universe are bound together in larger clusters. The Milky Way is part of a collection of about fifty-four galaxies called the Local Group, which is itself part of a much larger Virgo Supercluster. There's an even larger supercluster which encompasses Virgo and other superclusters into a massive association called Laniakea.

The study of clusters and superclusters gets astronomers into a realm of research which focuses on what's called the 'large-scale structure' of the universe. It's a hierarchical arrangement and begins with the smaller units (such as the Local Group) and expands outwards to the largest superclusters arranged in what looks like a cosmic network of matter. This is where the search for the mysterious 'dark matter' is important and observatories are being used to measure its effects on galaxies and clusters.

About a decade after Hubble's observation, astronomer and theoretician Fritz Zwicky was observing galaxies in the Coma Cluster, which contains about a thousand galaxies. Astronomers had noted some strange gravitational anomalies in the cluster's motions and Zwicky's observations indicated that its galaxies were moving faster than they should be if they were made solely of visible matter. He suspected the existence of something which couldn't be seen visually, but still had an effect on the galaxy. He called it 'dunkle materie', German for 'dark matter'.

In the 1970s, astronomer Vera Rubin and her colleagues were observing motions of stars in galaxies and noticed those 'kinematics' also indicated something 'invisible'. Their observations proved the existence of this dark matter. Observatories are playing a major role in helping astronomers understand more about this hidden 'stuff', its contributions to galaxies, and its distribution throughout the universe.

Oddities in the Cosmic Zoo

Observatories are often called upon to focus on high-energy transient events in the universe. They include supernova explosions, of course, as well as such cosmic oddities as pulsars, quasars, and gamma-ray bursts. Pulsars are rapidly spinning neutron stars which are usually the result of the death of a massive star in a supernova explosion. The first one was discovered in 1967 by Dame Jocelyn Bell Burnell, as discussed earlier. The 'signal' from this pulsar was easily detectable in the radio part of the electromagnetic spectrum. The term 'pulsar' comes from a mashing together of the words 'pulsing' and 'star'. Since that time, many more pulsars have been detected and observed.

Quasars are also radio-bright, but are much, much larger than pulsars. A quasar, short for quasi-stellar radio source (also called QSO for quasi-stellar object), is a very bright active galactic nucleus. They're caused by gas and other material falling onto the accretion disk surrounding the central supermassive black hole in a galaxy.

As the material swirls into the disk, it gets superheated and gives off a huge amount of energy that is released as light across the electromagnetic spectrum. So, while quasars are radio bright, they may well also have a visible light component, and give off x-rays and other radiation. These are objects which can be studied by a variety of telescopes, and each different part of the spectrum in which they radiate tells a different part of the total story about quasars.

Gamma-ray bursts (GRBs) have long posed a challenge to astronomers, not just to explain them, but also to catch them in the act. They are exactly what the name says: a very fast, bright burst of gamma-rays from a distant galaxy. They can be as short as a few milliseconds (tiny fractions of a second) to lengthy outbursts which can extend over a few hours. After the initial intense emissions, the object exhibits an afterglow visible in other wavelengths and frequencies. So, what are these insanely bright events?

The best theory to explain the incredibly high energy involved in a GRB suggests that there are two types: long-duration and short-duration. The short ones are most likely caused when two neutron stars merge to form a new black hole. Long-duration bursts can happen when a supermassive star dies and collapses to form a black hole.

The arrival of a GRB signal on Earth is a cue for observatories to spring into action to catch the fading afterglow in other wavelengths. These are the most violent explosions in the universe, and astronomers are still working very hard to observe them as they happen.

The Big Bang

One of the most challenging questions in astronomy is 'how did it all get started?' The study of the origin and evolution of the universe is a specialty called 'cosmology' and it is continually looking for ways to look further back in time. The point where the universe began is called 'The Big Bang', a term coined by the late British astronomer Sir Fred Hoyle. It is a cosmological model that describes the universe starting as a high-density and high-temperature state in a tiny point (a singularity) and expanding out to become what we observe today. Unfortunately, no telescope can see back to the Big Bang to confirm exactly what happened. Astronomers can, however, see the 'imprint' of the Big Bang's energies in something called the 'cosmic microwave background radiation' (CMBR). That was first seen in observations made by two scientists, Arno Penzias and Robert Wilson, who were doing microwave observations and detected a uniform microwave glow in the sky.

What is the CMBR?

In the young universe, there was a period right after the Big Bang when the infant universe was dense, hot, and filled with a glowing hot fog of hydrogen plasma. As it all cooled, the universe became more transparent, after which, photons of light could travel more freely through the young, expanding universe. Those photons have been traveling across space as it expands, and we see them today as the cosmic microwave background. Although this background emission was predicted, starting in the 1940s, it wasn't until Penzias and Wilson announced their work that the CMBR was confirmed. Since then, at least two orbiting telescopes – called the Cosmic Background Explorer (COBE) and the Wilkinson Microwave Anisotropy Probe – were used to make maps of the emissions. They haven't been the only observatories to focus on this area of study. Balloons have carried instruments up to high in the atmosphere to study it, and there have been various arrays built to study it from Earth, too.

The Early Universe

Beyond studies of this first light after the Big Bang, astronomers have been pushing their telescopes to study the first stars and galaxies that formed. The light from the first stars was largely ultraviolet, but as it has travelled across the expanding universe, its wavelengths have been shifted into the infrared. So, astronomers survey the sky using infrared, microwave, and low-frequency radio telescopes to search out those first stars.

The earliest stars began forming a few hundred million years after the Big Bang. As the universe began to cool, gravitational attraction between clumps of hydrogen would pull them together. It took a while, but eventually the clumping resulted in dense clouds of gas with warm cores. Those cores were the central seeds of stars, where hydrogen could be fused into helium. The densest regions in the new universe became hotbeds of star birth activity, pumping out huge amounts of ultraviolet radiation which we now see in infrared wavelengths.

At the same time, astronomers also are studying the large-scale structure of the young universe. It looks like a delicate web of dense regions of stars connected by thin filaments of matter, and voids in between. Eventually, galaxies of stars formed at the knots in the web. Astronomers are also interested in the role both dark matter and dark energy played in the past and how it affects the present-day universe. While dark matter cannot be directly seen, its gravity affects the matter we can see (called baryonic matter). This gravitational 'presence' can be detected by an effect called gravitational lensing. That's the distortion introduced in light as it passes by very large gravitational fields. Since dark matter has a gravitational pull, it combines with the mass of objects such as galaxies and galaxy clusters to distort light. Several observatories, such as Subaru and Gemini, are studying this lensing across the universe. New facilities are also coming online to probe the essence of dark matter and dark energy more directly.

Dark energy exists as part of the universe – the largest part. A pie chart of all the known 'stuff' in the universe would show a very tiny slice that represents baryonic matter, the planets, stars, galaxies, and so on. It's about 4.6 per cent of the total mass. Dark matter takes up about 27 per cent, and dark energy is a whopping 68 per cent, which is a lot of 'something' out there. And, astronomers still aren't sure what it is. They know it's a form of energy that is spread throughout space, and its main influence seems to be accelerating the expansion of the universe. It's a property of space, but beyond that, the ideas about what it actually is remain under discussion and the subject of research.

Astronomy Shows It All

We have always lived in a multi-wavelength universe our stargazing ancestors would find amazing in the extreme. Today we are fortunate to be able to observe it with the latest technology and astronomers have an astonishing wealth of objects to observe, each interconnected with their galaxies, and galaxies connected by streams of matter. While it may seem as if they observe multitudes of different 'things', in the end, what astronomers are finding are chains of events that result in objects which are themselves evolving as we watch. So, for example, a study of the Orion Nebula may seem at first like a study of the births of stars. But, it's also an examination of clouds of gas and dust. Embedded within those clouds are even tighter, denser constructs which turn out to be baby solar systems in formation. Those clouds will light up with star death as their most massive stars start to die in a few tens of millions of years. And, perhaps, someday, another young astronomer will find a pulsar blazing away after a supernova explosion. Another one will spend his or her career charting the mix of hydrogen, helium, carbon, nitrogen, and other gases – and molecules – swirling about in the nebula itself.

That's just one nebula in one galaxy. There are countless others to study, each one offering visions of star birth, star death, black hole activity, and much, much more. Observatories are our portal to all the universe has to offer. We have built them throughout our history, and most of them have stood the test of time, the ravages of nature and war, and the vagaries of funding and politics, to deliver continual and unstinting views of the cosmos.

2

THE EARLY HISTORY OF OBSERVATORIES

Sky watching is humanity's first science. Humans have a long and storied history of looking up at the sky and wondering what all those things are 'out there'. Chances are, it wasn't too long before people noticed a connection between seasons and time and the stars. Whoever did it first planted the seeds of astronomy. In the past, astronomers had only their eyes, but they were enough to get us started on the road to knowing our place in the universe. Today's astronomers continue in the footsteps of the earliest skygazers, but now they use mammoth automated telescopes and arrays, some ground-based, some space-based. They take advantage of computerised data-gathering and extended analysis to understand what's 'out there'.

It's always interesting to think about the first stargazers – those who stood outside their caves or lodges or fire circles to observe the sky. As it turns out, planet Earth is dotted with places where people watched the Sun, Moon, planets, and stars. For most of our history, sky gazing was done with the unaided eye in dark-sky conditions. It's a heritage of observation which enabled timekeeping, navigation, and ultimately, survival. That's a part of humanity's heritage rapidly disappearing in the modern age, with the rise of light pollution.

In the most recent centuries, with the rise of scientific research of the sky, observational tools grew ever more complex and technical. The history of the observatory, at least since the 1600s, has been dominated by the rise of one instrument: the telescope, in all its many forms. Today, astrophysicists use a range of instruments and detectors to make their discoveries, including some equipment that bears no resemblance to a telescope. Yet, all of it gets used to explore the universe.

Before we move on to the telescope and its evolution later in this chapter, a condensed history of observational astronomy gives insight into the practices of stargazing before the rise of modern observatories. We are, after all, descendants of stargazers who learned to use the sky for practical as well as ritual purposes.

Prehistory

Early observation posts and temples date back far into prehistory and in a few places the ancients left us images of what they saw. Some researchers point to the famous cave paintings at Lascaux, France, as an example of early 'space art' in the service of observation. In those darkened, long-hidden caverns, artists who lived some 15,000 to 17,000 years ago reproduced what they saw in the night sky. There are very distinct drawings of the Pleiades and Hyades star clusters, in addition to depictions of animals associated with them. In this case, they painted an early view of the constellation we now call Taurus the Bull.

The meanings of these paintings are lost to us through time's passing. However, scholars suspect they could be anything from natural calendars to some form of all-sky 'cosmography' – that is, an attempt to understand the nature of our planet and the rest of the universe. For scholars of archaeoastronomy (the astronomy of the ancients), the existence of the Lascaux paintings and their thought-provoking subject matter makes an interesting case for pre-Stone Age human interest in the sky. And, the prehistoric humans who made those drawings were not alone in their observations.

In Australia, Aboriginal sites exist which appear to be related to stargazing and date back well over 11,000 years. At least one area, called the Wurdi Youang arrangement, appears to be related to the rising and setting points of the Sun throughout the year. Why would early people build such a stone circle? It makes sense if the builders were not nomadic but were more settled, perhaps hunter-gatherers who also grew some of their food supplies. Knowing the times of the solstices and equinoxes would give them a kind of rudimentary calendar to guide them in planting and harvesting. According to some scholars of Aboriginal practices, even these earliest civilizations had sophisticated knowledge of the motions of the Sun, Moon, and planets, as well as a good understanding of the slow seasonal change of the stars overhead.

Of course, early people had no special tools beyond their naked-eye view of the sky. However, even a year's worth of sky watching likely gave them a rudimentary understanding of the apparent motions of sky objects. Associating those changes with the seasons was the next step toward connecting Earth and sky for practical purposes. It's very likely that hunting, planting, gathering, harvest, and navigation were all activities driven by seasonal change. And, what better way to document the change of seasons than by using the sky as a calendar? Other connections, including the religious and civic, probably followed in short order. Over time, this human 'use' of the sky led to the establishment of specific observing places, the predecessors to our modern observatories.

The earliest known evidence for dedicated places we could call 'observatories' dates back to a period at least some 3,500 years BCE, sometime around the late Neolithic Stone Age and the start of the Bronze Age. For example, early Egyptians of the Early Dynastic Period (3150 to 2613 BCE) built sundials into larger structures such as obelisks to help keep track of time. Babylonian astronomy stretches back to around 3500 BCE and they were the first to record their observations on clay disks.

China has had a long history of astronomical observation, dating back to several thousand years before the Common Era. As with most other cultures, it's likely the ancient Chinese first began recording their observations of the sky as a way of noting the passage of time and the seasons. Their early calendars were based on phases of the Moon, with some adjustments to account for extra days in the year. Around 2300 BCE, scholars and skywatchers at Taosi in northern China built an observation site with a gnomon to help them keep track of celestial objects by using shadow measurements (much the same as modern sundials do today). This site is the oldest known astronomical observatory in Asia.

Such measurements were important for calendar-making and each dynasty had a calendar as its symbol. In many senses, knowledge of the sky, of timekeeping, and calendar-making, was connected to political and religious power throughout early cultures, and was no different in China. Ancient traditions in the country dictated that the kings and emperors were 'blessed' by the heavens, getting their political power from the perceived motions of objects in the sky. It was almost as if the sky conferred a kind of 'divinity' on the elite. It also made astronomers and the astrological predictions they made based on the positions of the stars and planets about as powerful as the rulers they served. This was true across many ancient cultures.

As other civilizations did, the Chinese created their own constellations (star patterns). They divided the sky into 28 lunar mansions, and had a number of smaller patterns or asterisms. There's some evidence of Babylonian and Indian influence on the Chinese system of star patterns, but for the most part, their constellations were developed independently, and don't necessarily correlate with the Greek and Roman star patterns we know of today. The oldest of the constellations recorded on star maps date back to around the 7th century, when the Tang dynasty began its rise to power.

Not only did the Chinese chart star positions and patterns, but they began observing and recording lunar and solar eclipses starting in 750 BCE. They also noted the positions of 'guest stars' (supernovae and novae), and the changing positions of the planets.

China is also identified with the earliest development of celestial globes, the most familiar of which is the armillary sphere. This is a specialised celestial globe first invented around the 4th century BCE. It, and other globes like it, were used to measure north polar distance (or declination) and from that measurement, calculate right ascension. These are coordinates which help astronomers pinpoint the position of a star or other object in the sky. They describe a location in the sky in much the same way that latitude and longitude do for locations here on Earth. For a given sky object right ascension and declination are its celestial coordinates. Declination is the same as an object's latitude and right ascension is the same as its longitude.

Today, these are abbreviated as DEC and RA. So, for example, the North Star (Polaris) is RA 02h 31m 48.7s and DEC is +89 degrees, 15' 51". They are expressed in hours, degrees, minutes and seconds of angular measure.

Chinese astronomy influenced nearby Japan, conveying information about calendar-making and making similar assumptions about the stars' influence on Earthly human affairs. Astronomers in Japan were, like their counterparts in China, expected to predict sky events such as eclipses. The Japanese constellations were, like the Chinese, divided into houses, with each house linked to a specific town or region on the islands. Celestial events that occurred against that backdrop were either good or bad, depending on how they were interpreted.

Egyptians and the Stars

The ancient Egyptians had a very intimate relationship with the stars. Why was this so? Partly for mystical reasons, but also for some very practical ones. The civilization along the northern part of Nile depended on the river and its annual flood season. The resulting deluge enriched the lands on either side of the river, enabling the Egyptians to grow their crops and feed themselves. Knowing exactly when the flooding would start was of paramount of importance. As other early civilizations did, the early Egyptians developed a calendar and methods of timekeeping. They used the sky for this purpose, noting the rising and setting times of the Sun, Moon, and various stars. Their observations were used to predict the onset of flooding.

Thus, Egyptian astronomers watched for the heliacal rising of the star Sirius each year. That is the time when it can first be glimpsed rising just before the Sun. It marked not just the beginning of flood season, but also the start of the new year. About 3000 BCE, this rising occurred in early July as seen from the ancient city of Memphis along the Nile. (Today, this rising occurs in August, due to a phenomenon called the 'precession of the equinoxes'.)

The flooding of the Nile wasn't the only thing that ancient astronomers associated with stars in the sky. The most famous monuments of the ancient world, the three pyramids at Giza in Egypt have very distinctive ties to certain stars in the sky. First, they appear to nearly in alignment with the cardinal points: north, south, east, and west, varying only by a small amount. The small difference of a few degrees off from true north makes sense if the Egyptians aligned to what was then their pole star.

Today, our pole star is Polaris (also known as Alpha Ursa Minoris), but back in their time, the star everyone looked at to determine north was a relatively dim one called 10 Draconis. Again, this shift is due to the precession of Earth on its axis. The pyramid builders also used other non-stellar methods to align and build these monuments, but the connection to the sky was undeniable. The pyramids are located on the west side of the Nile, a direction that was always associated with death. Scholars of archaeoastronomy know of other connections to the stars within the pyramids, as well, and it's very clear the Nile civilization was heavily attuned to the sky for many of its ritual, calendrical, and architectural purposes.

In ancient Greece, philosophers and scholars regarded astronomy as a branch of mathematics and a kind of natural philosophy. They searched for some math-savvy way to explain the motions of the planets, using geometrical principles to explain those motions.

In ancient India, sky watchers used astronomical observations to create calendars, and developed methods for tracking the motions of the Sun and Moon. There was clearly some association with religious ritual, and scholars worked out computations for the length of the Earth day, and the periods of the naked-eye planets.

Observers began charting astronomical objects and events around 2000 BCE. They calculated eclipses, determined the circumference of our planet, theorised the Sun was a star and observed the planets. Like the Chinese and many others in the ancient world, Indian astronomers in the prehistoric era used the gnomon for shadow observations. Also like the Chinese, the Indians used the armillary sphere, in addition to other instruments for sun-sighting activities and investigation of polar motion. Also as in many other cultures, for Indians, astronomy was linked to the practice of astrology.

Henges

Astronomical interest wasn't limited to Europe, Asia and the Middle East. Cultures around the world had links to the sky. In Africa, for example, there's a desert basin called the Nabta Playa in southern Egypt. People who lived some 6,000–7,000 years ago erected stones as part of what appears to be a giant solar 'henge'. This calendar circle of stones seems to have specific points that align with the summer solstice, which would be the start of the area's rainy season. There was more than a ceremonial aspect to this spot. Archaeoastronomers suspect the area was used for large seasonal gatherings, and in later centuries, the stone circle (and later monuments) were likely built to connect the nomadic people of the region to Sun and stars in a kind of calendrical way. This ancient observatory was recently discovered by archaeologists, and is sometimes referred to as 'Africa's Stonehenge'.

Although not as old as the African henge, the Big Horn Medicine Wheel in the Big Horn Mountains of Wyoming in the United States, appears to have a very direct link to events in the sky. The arrangement of stones seems to have been used as 'pointers' from this high-altitude mountaintop to predict the rising and setting places of the Sun and the stars Sirius, Aldebaran, and Rigel. The wheel was constructed around the year 1200 CE, and was likely used for ritual and other purposes by the native Plains people who lived in the region at the time.

Probably one of the most famous of the prehistoric monuments connected in some way to objects in the sky lies in southern England. Stonehenge is mainly a burial site, first constructed in the Neolithic and Bronze ages, and used in some way by early societies. It is not really an observatory in the typical sense, but it does have a definite connection between ritual and the observation of celestial objects. Part of the monument aligns with the direction of specific midwinter and midsummer sunrise and sunset times. For example, one stone in the monument, called the Heel Stone, lines up approximately with the summer solstice sunrise to an observer standing at a certain spot. Experts continue to study Stonehenge for archaeological purposes, but it's very clear the builders and users of this still-standing circle of stones were quite familiar with the sky. They evidently assigned some connection between sunrises, sunsets, and the rituals of burial. To make this connection, they had to be proficient sky watchers.

A Celestial Passageway to the Afterlife

In Ireland, the Megalithic Passage Tomb at Newgrange was built about 3200 BCE is one of the earliest structures known clearly oriented to specific astronomical events. The interior passage and inner chamber are illuminated each year by the winter solstice sunrise, and for a few days either side of it.

The natives of the western hemisphere also watched the Sun, Moon, planets and stars. Their interests were also calendrical, and in many cases, such observations were connected to important rituals. Their observing sites are scattered throughout North, South, and Central America, and include everything from elaborate temples to simple circles of stone or wood laid out to mark important celestial rising and setting points. Clearly the ancients came to an early understanding of the connection between time and the apparent motions of sky objects, and adopted them for their own ritual, calendrical, and timekeeping uses.

Polynesian Wayfaring by the Stars

Stargazing was not limited to the land-bound in ancient times. Most readers know of the Phoenicians, who spread across the Mediterranean beginning around 1500 BCE. They were among many in the ancient world who mastered the use of stellar positions (and rising and setting) to help them navigate. The Minoans, Greeks, and various Asian cultures did the same. What about people who were entirely surrounded by ocean? That was the case for Polynesian Islanders. Like other ancient cultures, they, too, had no telescopes or observatories, but were skilful observers of the sky. They developed some instruments to help them, but their 'toolkit' for navigation also included extensive knowledge of the seas and passed the information down through the generations. In particular, navigation by the stars required them to know the positions of the stars at various times of the year, as well as their rising and setting times. Using their knowledge, the Polynesians explored up and down the Pacific Ocean, navigating huge distances across open ocean.

Astronomy of the Maya

Some cultures elevated their use of astronomy for calendrical and ritual purposes to a high art. Among them were the Pre-Columbian civilizations in Central America (the region connecting North and South America) and Mexico. The Mayans and the Aztec are two of the most famous examples. The Maya inhabitants of the region began creating their civilization around 2000 BCE. Their astronomical observations were largely connected to astrological predictions and not out of any sense of 'doing science' to understand the objects in the sky.

While they built no observatories like those astronomers use today, the Maya did create elaborate temples from which their astronomer-priests and priestesses watched the sky. The ancient Maya developed their own writing and numbering systems, and their calendars were quite complex. They had a refined understanding of astronomical phenomena and, for example, had a very accurate estimate of the length of a month, and had a better idea of the length of a year than their European conquerors did. They also charted planetary motions and devised eclipse predictions, which required years of painstaking observations to achieve. For the Maya, as with other early civilizations, knowledge of the sky was not only important for timekeeping and calendar keeping, but it served as a form of power. The Mayan kings used the arcane knowledge of the stars to make important decisions about wars, marriages, rituals, and other aspects of their society. The Maya created a complex calendar for their rituals. Because of their location in the tropics, observations of sky objects when they crossed the zenith were important. There's quite a bit of evidence that the Sun's crossing the zenith on 26 July each year was used as the start of the Mayan new year. At least one temple contains a stone calendar commemorating this date for all posterity

Much of the Mayan cultural use of astronomy was passed on to other cultures, such as the Aztec, who also used celestial knowledge for their own ritual and social purposes. With the advent of European incursions from Spain, and the imposition of Catholic faith on the natives of the region, some of the most useful Mayan materials were destroyed. A few remain, including several manuscripts which encoded a record of solar eclipse and the motions of the planet Venus. In addition, some of their observatories remain as tourist attractions.

The Aztec people of central Mexico also had long experience observing the sky, mainly for calendrical and timekeeping purposes. Their calendar was incredibly complex, but made sense of their observations. It was based on two cycles: a 365-day cycle called *xiuhpohualli* and second on, based on a second 260-day one named *tonalpohualli*. Together, these formed a 52-year 'calendar round'. The *xiuhpohualli* is considered to be the agricultural calendar, since it is based on the Sun, and helped the farmers keep track of planting and harvest dates. The *tonalpohualli* was more of a ritual calendar for sacred ceremonies.

What was the basis for these complex calendar calculations? The Sun's apparent path across the sky was part of it, but there's evidence to support a prehistorical interest in the first pre-dawn appearance of a star cluster they called *Tianquiztli*. We know it today as the Pleiades. Over time, the Aztec began using the solstice and equinox sunrises as markers for their calendrical time. The most obvious artefact of their calendrical interest is evident on a giant stone piece called the Stone of the Sun, or *Tonatiuh*. Symbols carved on its surface indicated days and years in their calendar rounds. It was an incredibly intricate piece of work and reflected years of experienced observation by Aztec astronomers.

The Arcane Roots of Scientific Astronomy

As we've seen above in selected examples from prehistory, astronomy's earliest roots were less scientific and more founded in cultural practices. Until the European Renaissance, astronomy was largely a means to capturing and managing time and the passage of seasons, often connected to ritual and civic practices. Of course, as most astronomers become painfully aware when they meet members of the public, astronomy remained connected to astrology for centuries. For some people in the modern age, it still is. Eventually, astronomy diverged from this pseudo-scientific pastime in the 18th century. Before then, however, astronomy and astrology were basically considered one and the same. For centuries, astronomy, astrology, religion, and political power were also tied together and it is worth taking a brief look at how that worked.

Astronomy in the Service of Ritual

The union of the religious ritual and the secular observation of the sky wasn't just confined to early human cultures. Certainly there was a connection between the sky and ritual in many places throughout the world. While the science of astronomy has some roots in ancient religions and myths, it's easy to understand why. Our earliest ancestors really had no way of understanding the actual nature of things they saw in the sky. They could, at best, measure motions, watch the slow change of the sky over the seasons, and chart everything they saw.

Western astronomy has many roots in the Fertile Crescent of Mesopotamia, dating back at least a thousand years before the Common Era (if not earlier). The cultures there developed writing and recordkeeping, and there's quite a lot of evidence that Mesopotamian theology included references to planetary gods. At least in Old Babylonia, astrologer-priests began to apply mathematics to calculate and document the changes in the length of a day over a year, and they also recorded planet and star positions throughout the year. The Babylonians also created the MUL.APIN, a catalogue or compendium of stars and constellations, along with information about their rising and setting times.

In later centuries, particularly in the early years of the Common Era, pre-Islamic Arab observers were avid sky watchers, usually charting the rise and set times of the Sun, and certain stars. As the Islamic religion spread, mathematical calculations became part of the astronomer's tool kit.

Middle Eastern scholars not only assimilated and continued to refine much star lore and astronomical knowledge, particularly from the Greeks and Egyptians, as well as Indian and Chinese knowledge. They also began developing (if not inventing) instruments such as the astrolabe. They also created observatories for the express purpose of tracking stars and the Moon's appearance in the sky. That latter is an important marker in Islamic religion, helping its adherents know such things as the start of worship, the beginning and ending of Ramadan, and other important ritual dates.

During the Islamic Caliphate period – the 9th to 13th centuries CE, sometimes referred to as the Golden Age of Islam – Islamic scholars brought their observational practices and refinements of such things as Ptolemaic theory throughout the lands they occupied. (The Greek-Roman scholar Ptolemy believed the Earth was the centre of the universe.) The Islamic scholars built a number of observatories throughout the Mediterranean and Middle East, equipped with meridians, astrolabes and other equipment. There, observers measured and recorded the motions of the Sun, Moon, and planets. Islamic astronomers published their observations of celestial objects in addition to their translations of works from the ancient world. The first was the *Zij al-Sindh*, written by Muhammed Mūsā al-Khwārizmī in the year 830. It was a compilation of tables showing

the motions of the Sun, Moon, and the five naked-eye planets –
Mercury, Venus, Mars, Jupiter, and Saturn. His work reflects the
Ptolemaic world view and introduced methods for calculating tables
predicting such motions. Other scholars and works appeared in
succeeding centuries, with some delving deep into the Ptolemaic
view of an Earth-cantered universe, and others disagreeing with the
notion.

Throughout the Dark Ages in Europe, after the fall of the Roman
Empire, and the rise of Christianity throughout the Continent, certain
information from ancient Greece was considered paganistic and not
consistent with Church teachings. Against this backdrop, Islamic
scholars safeguarded the knowledge of the skies they had gathered
from others and made on their own, and eventually passed it along as
part of their conquests.

During the medieval centuries of the three caliphates, Arabic
scholars translated much of the ancient Greek knowledge of the sky,
mathematics, geography, and navigation into the Arabic language.
This not only served to educate the Islamic Muslim world, but
also preserved the knowledge. Later on, translations into other
languages helped spread it throughout Europe. The caliphate in
Spain established extensive libraries in cities throughout the region
they called Al-Andalus and the region was dotted with observatories.
In Baghdad, Cordoba, and Toledo, for example, libraries boasted
hundreds of thousands of books, and these places didn't exist solely
for use by Islamic scholars. Starting in the late Middle Ages, Muslim
science became a commodity, with people traveling from other parts
of the world to learn about science, mathematics, and other topics.

As the Islamic influence began to wane and Christianity spread,
the libraries in Spain, Italy, and other places came under the rule
of the dominant Church. People began translating the Arab texts,
and within a century much of the lost knowledge from the ancient
world had been recovered. This included treatises on astronomy,
medical science, and philosophy. Thanks to their scholarship, we
see Islamic and Arabic influences in astronomy and mathematics,
including our numbering system (called the Arabic numerals) and in
many star names, such as Betelgeuse (*Yad al-Jauzā'* which means 'the
hand of Orion') and Aldebaran (Arabic *al-Dabarān*, which means
'the follower').

Astronomy and Early Christianity

Even the early Catholic Church maintained a connection with astronomy that traces back to at least medieval times. Like other religions, Catholicism (and later on, other Christian faiths) needed astronomical information to help determine such dates as the occurrence of Easter for the purposes of celebrating the proper rituals. It was an entirely proper way to use observations of the sky: for an accurate understanding of seasons in order to fix important dates on a liturgical calendar. At least one church building, the Basilica of San Petronio in Bologna, contains a good example of what's called a 'meridian' line. They also exist in other churches, specifically in Rome, Florence, Paris, and Durham Cathedral in England.

What is a meridian line? It's basically something that measures the passage of a celestial object over the highest point in the sky. In the case of the churches or – for example, the Gottlieb Transit Corridor located at Griffith Observatory in Los Angeles, California – such a meridian 'instrument' measures the transit of the Sun over a north–south line. Over the course of a year, the Sun appears to move north in the sky during parts of the year and south for the rest of the year. As it goes through this apparent motion (and it's not really the Sun moving, it's really Earth spinning on its axis, which is tilted), a ray of light from the Sun appears to shifts along the floor where such a meridian line is placed.

The farthest north point on the line (in the northern hemisphere) is the summer (or June) solstice. The furthest south point is the winter (or December) solstice. The midpoint is the equinox. The date of Easter is determined by a fairly complex set of rules which call for it to be celebrated on the first Sunday after the first Full Moon after the spring equinox (which occurs on 20 March in the northern hemisphere each year). By placing an accurately sited meridian line on the floor of a church (which would give plenty of room for such an astronomical instrument), the clergy and church hierarchy assured themselves of a constant way of determining an important ritual date. In this sense, the Catholic Church continued a tradition practiced by many other cultures which relied on the sky for ritual purposes. All those 'date-determining' practices were not necessarily scientific in nature, but they set the stage for further exploration of the sky by later people.

Astronomy in the Service of Politics: Astrology

Throughout human history, rulers have gained political power through many different ways: conquests (usually wars), votes, kingships, and queenships. While stargazing isn't often thought of today as a means of political advance, in the past, in some societies, the stars played a significant role indeed. The very basis of astronomy's use of star positions and planetary motions is the practice of astrology. It's the belief that the positions of planets against the backdrop of stars somehow influence human behaviour and can be used to predict political success, love, and other pieces of the human condition. For centuries, astrology kept a great many astronomers employed, and in some places conferred elite power on them. Knowing the positions of the planets required meticulous observations and record-keeping, and as mentioned elsewhere, there is evidence of these kinds of tables in early Babylonia.

The Greek-Roman scholar Claudius Ptolemy, who suggested Earth was the centre of the universe, created a work called the *Tetrabiblos,* which was his treatise on the practices of astrology and how to use it for predictive purposes. It doesn't take much to imagine a wily astronomer telling his or her patron (a king or an emperor) that the stars looked favourable for an invasion, the execution of a rival, a marriage or some other action he wanted to take. Or, perhaps an astrologer would be called in to do a reading for a new spouse or a baby. Politics being what it is, it's very likely astrologers were also required to use their elite knowledge of the stars to figure out which days would be good to dedicate a temple or open a market or other civic show of power. Certainly, the Greeks and early Middle Eastern societies weren't the only ones to use astrology as a show of arcane knowledge supporting elite power. The Romans practised it too.

Up until the Renaissance, in fact, astrology was taught in universities alongside other sciences. It certainly required dedicated observations of the sky. Planetary motions and star-chart making were equally as important to astrology as they were to astronomers. Johannes Kepler worked very hard on trying to bring scientific rigor to astrology, even as he was developing his idea of a heliocentric universe.

Interestingly, for quite some time, astrology was an important part of medicine. There was the overall belief that the body and future of a new-born was influenced by the position of a planet against a constellation backdrop of stars overhead at the time of birth. That is still a tenet of modern astrology, even though the constellations have shifted and astronomers continually point out that the gravitational influence of the doctor assisting the baby out of the birth canal is many times stronger than the weak (almost non-existent) pull of a distant planet. Medieval astrology also held that the position of the Moon had to be favourable before a physician could practice bloodletting on a patient. Different signs of the Zodiac (the constellations through which the Sun, Moon and planets appear to move against in the sky) influenced different parts of the body. Proper astrological readings had to be done before any healing could commence.

In China, knowledge of the stars conferred an elite status on the court astronomers who really were astrologers practicing arcane divinations to foretell good and bad events. For example, the rulers from the Jin dynasty brought astronomical instruments to Beijing where they built the first observatory. It still exists today as the Beijing Ancient Observatory and preserves historical artefacts. When Kublai Khan assumed control in 1279, he built a new observatory nearby. His observatory was then moved to Nanjing by the new Ming emperor. The transfer of power from one man to another, one dynasty to the next, almost always seemed to convey astronomical knowledge and status. It wasn't until the mid-17th century that the observations of the sky moved beyond the simply astrological to true scientific study of the stars and planets. Today, the Ming-era instruments are housed as artifacts at the Purple Mountain Observatory in Nanjing.

While astrology is now known to be more of a parlour game and not much of a scientific study of the stars, its spread did require disciplined observing, accurate record-keeping, and eventually an intimate knowledge of Sun, Moon, and planets against the backdrop of stars. Observatories that were built for this practice, and to gather useful knowledge for navigation around the planet, were the forebears of the facilities we use today to explore the universe.

Early Attempts at a Scientific Understanding the Sky

Most readers are familiar with the connection between the sky and the myths and legends in the ancient world. The early Greek legends, the ancient mythologies of Egypt, the Middle East, India, China, and cultures in other parts of the world all had tales which connected people to the sky. People noticed the sky and what was up there, how objects appeared to move, and what celestial events occurred. The Greek philosopher and poet Homer, for example, and another writer named Hesiod, referred to well-known constellations and stars in their works, as well as eclipses and other celestial events. It's clear that both writers were already familiar with the sky and the cultural and religious associations it had.

Efforts to make some scientific sense of the sky came later, with people trying to solve astronomical problems using mathematical and geometrical models. Among the greatest of the Greek philosophers to concern themselves with astronomy were Hipparchus, who developed early star catalogues; Eratosthenes, who worked on measuring Earth; and Aristarchus, who first proposed the heliocentric model of the solar system. Perhaps the most famous of the Greeks was Claudius Ptolemaeus, also known as Claudius Ptolemy, a mathematician who observed the sky and catalogued what he saw. He also wrote treatises about astronomy and astrology.

None of these people had observatories as we know them today, but it's clear they were well-versed in observing the sky with the unaided eye. They might be more akin to amateur observers who, though not having formal training in astronomy, often know the sky and its celestial objects very well. It fell to much later astronomers to do the heavy lifting of postulating theories about celestial motions of objects to explain what appeared to the ancients as a somewhat 'clockwork' universe.

One such thinker was Polish astronomer Nicolaus Copernicus (1473–1543), who lived just prior to the invention of the telescope. He spent much of his effort developing his ideas that the planets orbited the Sun (as a fixed point in space), and that Earth turned on its axis daily. His 'heliocentric' system triggered a wave of scientific interest in the sun-centred universe, in what some called the 'Copernican Revolution'. Yet, even he did not have a magnified view of the sky. Copernicus was, like everyone up to the time of Galileo, a naked-eye observer. He was also a prodigious theoretician, developing explanations for planetary motions and cosmology. He published his theories and ideas in his work *De revolutionibus orbium coelestium*, which described his Sun-centred universe. Although it gathered little 'official' religious attention at the time, the Church (of which he was a part), later condemned his work as being against holy teaching. It even took the step of warning Italian astronomer Galileo Galilei against defending Copernican theories. However, Galileo couldn't stop the flow of scientific knowledge of the sky once he and others learned to magnify their view of celestial objects.

The Evolution of the Telescope

The telescope has been around since the early 17th century, and evolved from a simple tube in Galileo's time to the behemoth ground-based and space-based instruments used by astronomers today. In most people's minds, Galileo's name is inextricably linked with the telescope. This is because he was the first astronomer recorded to use optics to magnify his view of the heavens. Galileo first described his observations in a work called *Sidereus Nuncius (Starry Messenger)* published in 1610. In it, he told of his observations of celestial objects through a telescope of his own making.

There's a common idea that Galileo himself came up with the first telescope. He did build one, and many astronomy books show paintings of him at the eyepiece. However, he wasn't the inventor. He had heard about an optical spyglass and is now best known as an 'early adopter'. He built his own instrument based on technology that already existed by the time he set up observing on a hillside near Padua, Italy, in 1609. It was a tiny one, even by today's backyard-telescope standards. It was only 26 mm (11 inches) across and magnified his view by a power of three. Of course, it wasn't big enough, and he manifested what may be the first known case of 'aperture fever'. That's the desire for ever-larger viewing apertures, which means larger telescopes. It affects people even today.

Galileo began building better optics, jumping up to 20-power and eventually 30-power magnification. He also built scopes for others, selling them as spyglasses. To the betterment of astronomy, his new tool allowed Galileo to observe the Moon's mountains and craters. Better magnification allowed him to study the Milky Way and discover that it's full of thousands of stars. His most famous discovery, captured for posterity in his drawings, was of Jupiter when he spotted its four largest moons: Io, Europa, Ganymede, and Callisto.

Who Invented the Telescope?

Based on Galileo's use of and popularisation of the telescope, it's easy to see why people think he invented it. However, the credit for the work belongs to someone else, even though there are debates about who, exactly, it might be. The leading theory is that a Dutch eyeglass maker was the inventor. His name was Hans Lippershey (sometimes spelled 'Lipperhey'), and he originally created a tube with two lenses in it to magnify distant objects. The idea was to offer it to the military for use, particularly as a spyglass to note distant ships or armies on the move. While he may not have been the first to actually think of the idea, Lippershey jumped to apply for a patent for his idea in 1608. It's also possible that an English inventor named Leonard Digges was the 'father' of the telescope. Regardless of who actually created the first telescope, news of it and its cousin the spyglass eventually reached Galileo. He promptly built his own, and the rest is history.

Optics and Aperture Fever

The invention of the telescope set in motion a technological race for perfect viewing that continues to this day with ever-larger telescopes and space-based observatories. To get better views, however, the telescope needed to be improved, and other astronomers began working on upgrading the instrument. Johannes Kepler, an assistant to the great visual observer Tycho Brahe (more about him in the next chapter), was an astronomer, mathematician and astrologer. In his time, astronomy and astrology were still closely connected and Kepler made at least some of his living supplying astrological forecasts to his royal patrons. Kepler devoted much of his time to experiments in optics. He wanted to understand and improve the view through telescopes.

Ultimately, he designed and built one that came be known as the Keplerian telescope. It used two convex lenses which provided much higher magnification than the simple assembly that Galileo had used. They were refractors – they relied on the bending of light (known as refraction) to deliver a view of an object to the observer. The spherical shape of the primary lens leads to a condition called spherical aberration, which means the lens fails to focus light down to a sharp point. The results are blurry images. (Readers may recall the Hubble

Space Telescope suffered from the same condition, but for different reasons. We will talk more about that in Chapter 5.)

Early refractors suffered greatly from chromatic aberration, as well as spherical aberration. These are conditions which plague telescopes with poorly ground lenses or inferior quality glass. Chromatic aberration causes coloured 'fringes' around images of objects in a telescope. Technically speaking, chromatic aberration is nearly impossible to avoid, and astronomers today get around it by using what's called a 'doublet' or 'triplet' telescope design with two or three lenses in the optical path to mitigate the 'rainbow' colours around the edge of the view. Reflectors don't have the same problem.

Spherical aberration (which can affect reflectors, too), results from lenses that haven't been ground to focus the light to a sharp point. Refractors are still in use today, and benefit highly from great advances in grinding technology and better glass formulations.

Johannes Kepler's work improved on Galileo's design in many ways, although aberration was still a problem. The only other drawback was the Keplerian telescope's sharper views appeared 'upside-down' through the eyepiece. Galileo's telescope didn't invert the view, but it also had a very narrow field of view.

For astronomy to advance, however, better views afforded by improved optics were needed. It's one thing to look at an individual planet through a telescope and note its general characteristics. But observers generally want as much detail as they can get, such as clear and distinct views of specific details like the Great Red Spot on Jupiter. For observers to resolve individual stars in clusters or in the Milky Way, the optics had to be better.

However, larger optics (lenses), require larger aperture, hence larger telescopes. To create the sharpest views, astronomers began building larger and longer instruments which allow better resolution of fine details on a target. They also gather more light, which allows dimmer objects to be seen. Some, such as one built by Polish astronomer Johannes Hevelius in 1657, were up to 46 metres (150 feet) in length. This allowed them to sharpen the view, but at the cost of making their telescopes heavier and vastly more difficult to use.

In the realm of optics, astronomer Christiaan Huygens began working on his own telescope in Holland around the mid-1600s.

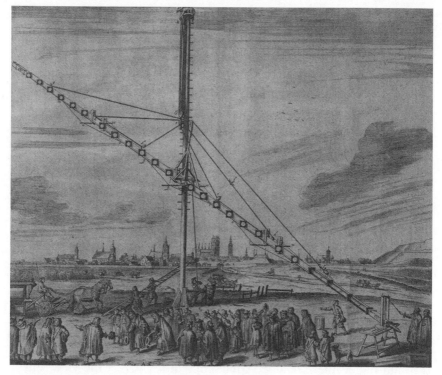

A woodcut illustration of a Keplerian telescope built by astronomer Johannes Hevelius. It was 46 metres (150 feet) long. Hevelius described building and using this instrument in his 1673 publication *Machina Coelestis*.

As a youth, Huygens (inspired by his father's interest in optics) began grinding his own lenses. Disappointed by the views through long telescopes and the difficulty of using them, he built a 3.7-metre-long (12-foot-long) telescope using his own optics and was able to make more detailed observations. His observations of Saturn, for example, allowed him to find a moon around the planet, and make more detailed studies of the planet's rings.

Hevelius, Huygens and others were grinding good lenses and building giant telescopes. But, they were unwieldy behemoths, difficult to operate and control. For example, on a windy night, a lengthy telescope would get blown around and be nearly useless for any kind of controlled viewing. Clearly, a shorter telescope was better for observing, but how would astronomers get good views?

The answer lay in mirrors.

Isaac Newton and the Reflecting Telescope

Visitors to Westminster Abbey can find tributes to many people, among them Sir Isaac Newton, who is buried there. He was a so-called 'natural philosopher' (a precursor to the modern scientist) who lived from 1642 to 1726. Newton was a highly accomplished man. He studied and invented concepts in mathematics (calculus), physics (mechanics and gravitation), and astronomy, and discovered the spectrum of visible light. He used prisms to study the nature of visible light, and applied his mathematical concepts to explain orbits of comets, the precession of the equinoxes. His three laws of motion are familiar to every student of physical sciences.

Of interest here is that he came up with an entirely new telescope design based on ideas developed earlier by Galileo and others. To accomplish this, he delved deeply into the physics of optics and found a way around the spherical and chromatic aberration problems that plagued refractors since Galileo's time was to build telescopes with reflecting mirrors. This led him to invent the Newtonian reflector using a copper and tin mirror as the reflector.

The design used a spherical mirror at the 'back' or 'bottom' of the telescope. Light from distant objects hits the mirror and is reflected to a secondary mirror that sends an image to the eyepiece (or an imaging device or sensor). It requires an exquisitely ground mirror and a shorter 'tube'. (The same 'reflecting' concept works in radio telescopes, too. However, instead of gathering light with a mirror, the radio telescope receives incoming radio signals and focuses them to a secondary 'feed' which then passes the data to a collection of instruments.)

Newton built his first telescope in 1668. It was a small one about 20 cm (8 inches) long. It had a concave primary mirror at the back end, which reflected light from a distant object to a secondary mirror. With it, he achieved the same magnification as refractors many times longer than his telescope. He largely solved the problem of chromatic aberration (since the light doesn't pass through a wedge-shaped piece of glass which refracts light like a prism does), but did not completely get rid of spherical aberration. For that, he needed more precisely ground mirrors. However, the technology to create those was just not available in his time. Newton's telescopes also faced another problem: tarnish. It reduced the amount of light the mirror could reflect, which was a problem refractors didn't have.

Based on Newton's work, telescope makers began building new instruments using larger spherical metal mirrors and experimented with a new shape called a paraboloid. The first person to actually create a working paraboloid mirror was English optician and mathematician John Hadley. His mirror didn't have the spherical aberration problem inherent in earlier designs. And, it meant astronomers could once again start building bigger telescopes that didn't have to be as long as the earlier ones. The size of the primary mirror became the indicator of how powerful the telescope was, rather than its length.

Throughout the 17th and 18th centuries, large metal-mirror telescopes were popular. Refractors were still in use, due to advancements in optics, but telescope makers found it less expensive to work with metal mirrors. However, reflectors still faced problems from tarnish and other environmental issues. And, as they got larger and larger, atmospheric conditions played havoc with even the best of the instruments.

The Royal Observatory at Greenwich, now part of the Royal Museums Greenwich in England, for a time boasted the largest reflector in the world. The Great Equatorial Telescope is a 71-cm (28-inch) instrument that was built in the late 1890s and commissioned in 1893. It was used to observe many objects, including double stars. Today, it is outfitted with cameras and computers for viewing. The actual observatory in Greenwich has a long history in both astronomy, as well as timekeeping and longitudinal calculations. Many visitors come to see its 'prime meridian' line that separates Western from Eastern hemisphere on Earth.

Artisans continued working with refractors. They included telescope maker Alvan Clark (and his sons), who used glass blanks made in England and Paris. From the 1840s onward, Alvan Clark & Sons ground lenses for such telescopes as the 47-cm (18.5-inch) instrument at Dearborn Observatory at the University of Chicago, plus telescopes for the US Naval Observatory in Washington DC, McCormick Observatory in Virginia, Pulkovo Observatory in Russia, and telescopes at Yerkes and Lowell Observatory in Arizona. Their smaller telescopes remain largely in the hands of collectors and clubs.

In England, Sir Howard Grubb headed up a family well-known for its large optical telescopes during the latter half of the 19th century. He founded the company Grubb Parsons, which made mirrors, mounts, and drives for telescopes in England and Ireland, Europe, Melbourne (Australia), North America, and Asia. It continued engineering telescopes until the 1980s, building instruments for the Anglo-Australian Telescope in Coonabarabran, Australia, The William Herschel Telescope and the Isaac Newton Telescope in the Canary Islands, and the UK Infrared Telescope on Mauna Kea, in Hawaii.

The largest professional (and last of the large observatory) refracting telescope is at Yerkes Observatory in Wisconsin. It was built in the late 1800s, and the building has only recently been closed. The telescope remained in use for more than 100 years.

Today, there are still refractors in use, particularly in the amateur arena. As mentioned above, modern refractors benefit from technological advances which have improved the glass and grinding process, and many amateurs prize those instruments for their quality views. The Skyscrapers Club of Rhode Island, for example, maintains an 8-inch Alvan Clark refractor that is still in use today. We will take a further look at amateur use in Chapter 8.

Charles Messier and the Fuzzy Objects

The 18th Century was a time of great discoveries in astronomy. The rise of the telescope enabled such men as Charles Messier (who lived from 1730 to 1817) to 'cruise' the sky in search of objects to study. Messier was a French astronomer who began observing celestial objects as a teenager, inspired by the appearance of a comet in 1944 to spend much of his adult life hunting down more comets. His search for 'faint fuzzy objects' resulted not just in the discovery of 13 comets over the course of 25 years (from 1760 to 1785), but also in an amazing list of objects known as the Messier Catalogue. During his comet searches, he found (or catalogued, along with his assistant Pierre Méchain) 110 other objects, including star clusters and nebulae. The idea behind his catalogue was simple: identify all those things that aren't comets so future astronomers could concentrate on finding comets and not worry about the other faint fuzzies 'out there'.

Messier couldn't have done his work without telescopes, and over the course of his career, he used nearly a dozen instruments. Most of them were refractors (but also a few reflectors) and he compiled much of his Messier Catalogue using a 100-mm (4-inch) telescope from an area near what is now downtown Paris.

The Age of Herschel

Sir William Herschel came onto the astronomy scene in the late 1700s, around the same time as Messier. He was a musician who also pursued a passion for the stars. He and his sister Caroline Herschel began their observing using refractors, but quickly switched to creating their own parabolic mirrors and reflecting telescopes. Together they experimented with different alloys for the mirrors in an effort to reduce tarnish and increase the amount of light they could gather. The first telescope Herschel built was fairly small, only about 15 cm (6 inches) across and 2.1 meters (7 feet) long. It had a magnification of about 40x and allowed him to see such things as Saturn's rings fairly clearly. He also built a 6-metre-long (20-foot) telescope with a 45-cm (18-inch) mirror, followed by his largest one in 1789. It measured 48 cm (19 inches) across.

William and his sister Caroline built and ran 'Observatory House' in Slough, England. Unfortunately, it no longer exists – although a commemorative plaque on the street where it once stood tells the story of one of astronomy's giants.

In 1785, Herschel and his sister began construction of what became known as the 'Great Forty-foot Telescope, so named for its length. For five decades, it was the largest telescope in the world, with two 120-cm (3.9-foot) concave metal mirrors enclosed inside an iron tube. Once it was nearly finished, Herschel set it up on the grounds of Observatory House and began making observations. Among other things, this may have been the telescope he used to discover the Saturnian moons Enceladus and Mimas. The telescope, although larger than his others, was difficult to use and often unwieldy. The Herschels often faced spates of bad weather, which limited use of this behemoth. Despite that, they used it until 1815.

Herschel and his sister are best known for their first sky survey, which they spent nearly a decade completing. The main search was for double stars, but along the way they catalogued clusters as well. On 13 March 1781, Herschel made his discovery of Uranus (using the large telescope). After it was verified, he gained the position as Court Astronomer to King George III.

Herschel's studies didn't stop at music and astronomy. As mentioned in Chapter 1, he also discovered infrared radiation (the section of the electromagnetic spectrum next to visible light), and used prisms and thermometers to measure the wavelengths of light radiating from stars. His studies of stellar spectra heralded the eventual advance of astronomy from a purely observational and cataloguing science to one which used physics to explain the characteristics of stars.

He also studied Mars and its seasonally changing ice caps, found the moons Titania and Oberon orbiting Uranus, as well as the moons Enceladus and Mimas around Saturn. It's safe to say the success of Herschel's work, along with that of his sister and his son John, was largely due to improvements in telescopes. Of course, it boded well for the spread of scientific astronomy around the globe.

The work of Herschel and others highlighted a rise in the study of astronomy and other sciences throughout the 18th Century. The scientists in this Age of Enlightenment were starting to test theories by experimentation and observation. Astronomy itself was moving in that direction, although it still supplied very practical applications for timekeeping and navigation. Trade with distant cultures was on the rise, along with the expansion of the British Empire, the colonisation of the 'New World' (and the eventual breakaway of the United States from Britain).

Observational astronomy now had Newton's theories of planetary and cometary motions to help explain their motions. Just after mid-century, the voyages of explorers such as Captain Cook kindled interest in the skies in the newly explored parts of the globe, particularly in the southern hemisphere. He took an expedition of naturalists and astronomers to witness the 1769 transit of Venus from Tahiti. Their work was of primary importance to navigators, particularly because they worked to calculate the longitude of their observing location. Finally, Herschel's discovery of the planet Uranus galvanised public interest. These events and voyages were instrumental in decisions to build observatories around the globe.

Hoisting a Leviathan

A nobleman by the name of William Parsons (the 3rd Earl of Rosse) was inspired by Herschel and his own desire to build the biggest and best telescope in the world. This 'Leviathan of Parsonstown', as it came to be known, was completed in 1845 on the grounds of Birr Castle, outside Dublin, in Ireland. Rosse had to learn mirror making on his own, since there were few records left from Herschel and others. He made a series of successively larger metal mirrors, culminating in a 1.8-metre (72-inch) mirror that became the heart of the Leviathan. The telescope's tube was hung between two stone walls, and used ropes and pulleys for deployment. Rosse kept a replacement mirror to insert into the tube when tarnish became an issue. From 1845 to 1917, this was the world's largest telescope.

The Leviathan was limited in its movement, which meant it could only really move 'up and down', and not much from side to side. This was fairly typical of telescope mounts up to that time and greatly affected how much sky the instrument could really 'see'. In addition, the area of Ireland where the telescope was built suffered from cloudiness and nasty weather, which introduced a new concept: atmospheric aberration. That's a condition caused by movements of air masses that cause the view through the telescope to look wavery and sometimes indistinct. However, despite the problems, Rosse's work showed larger mirrors could be made for astronomical observations.

Heading Down Under

As the great age of exploration that began in the mid-1400s continued through the centuries, of course astronomers became interested in skies other than those visible from the northern hemisphere. With this in mind, people making their way to such southern hemisphere countries as Australia and South Africa, and to South America, wanted to build telescopes and observatories there, as well. Sometimes these astronomy outposts were part of larger institutions in England and Europe but at other times, locals stepped in to build their own facilities.

For example, Australia's first large telescope, called the Melbourne Reflector, was built in 1869 to magnify the views of nebulae and star clusters in the southern skies. It was built by a committee of astronomers in Victoria and they chose a 121-cm (48-inch) metal mirror for the telescope. Unfortunately, just like its predecessors in the northern hemisphere, it was plagued by tarnish. To be recoated, it really needed to be sent back to Ireland, where it was originally made. Instead of doing that, the director of the observatory tried to polish the optics, but without any way to properly calibrate the job, the mirror was damaged. It didn't bode well for astronomers in Australia, and the telescope never did function well with the metal mirrors. Nonetheless, as we'll see later on, the telescope was used for many decades until it was sold, its mirror replaced, and the whole assembly moved to a different location.

The Silver Age

About the time Lord Rosse was building his telescopes, people had been experimenting with polished glass mirrors as possible substitutes for metal mirrors. Rather than put the silvering on the back of the mirror (as is common with regular mirrors we use for checking our appearance), people began putting finishes on the front of the 'cell'. In the mid-1850s, Justus von Liebig (a German chemist) found a way to cover a piece of ground glass with a silver film that theoretically would reflect more light than a simple polished piece of glass or metal. The silver was easy to work, but it was expensive and still could be affected by tarnish. However, it was much simpler to 're-silver' the mirror rather than re-polish out the tarnish from a piece of metal.

Before long, lightweight silvered glass cells found their way into telescopes. Those advancements, as well as improvements in grinding glass, paved the way for larger mirrors and the first of the 'Great Telescopes' of the late 19th and early 20th centuries. Advances in grinding provided the parabolic mirrors that are the hallmark of the reflecting telescopes today. Those mirrors, too, are coated with thin layers of silver and other metals to improve their reflectivity.

New Challenges

With the advances of highly polished silver parabolic mirrors, the stage was set for the creation of some truly large telescopes. The views through the improved instruments provided more magnified views of the cosmos, and more changes were on the way. In the 1800s, the invention of photography enabled astronomers to take actual images of the sky. Prior to that, observers were reduced to drawing what they saw and compiling voluminous written catalogues of positions and brightnesses. The additional application of spectroscopy (further discussed in the next chapter) added to the advances in the new age of scientific astronomy commencing late in the 19th century. There was one more piece of the puzzle to add: facing the challenge of finding good locations for telescopes to be placed. With the increasing use of electric lights and other forms of pollution, observatories had to leave population centres and move to the countryside and up in altitude. This, along with the rise of purpose-built laboratories for astronomy, first suggested in the 1800s, was also a hallmark of the coming age of Big Astronomy which flourished starting late in the 19th century.

3

THE SPREAD OF
OBSERVATORIES

From Islands to Mountaintops

The invention of the telescope and Galileo's first use of it for astronomy triggered a transformative age for astronomy, making it a science of discovery. It also set off a race to build bigger and better telescopes that continues to this day. In modern times, we live and operate in an era of very large telescopes, some measuring 8–10 metres (26–32 feet) in size. The largest optical/infrared one at the time of writing is about 40 metres across and will begin taking data in 2025. Even larger ones are on the drawing boards. Space observatories continue to extend our view of the cosmos from beyond Earth orbit and we will look at them in greater detail in Chapter 5. This chapter largely looks at the development of so-called 'modern observatories', particularly the ground-based facilities which appeared starting in the late 19th century and throughout the 20th. Along the way, we explore some fundamental changes that prepared observatories for the scientific role they were starting to play.

At the end of the 19th century, most (but not all) observatories were in the Northern Hemisphere. Even by the 1960s, a map showing the locations of the world's large professional facilities was still weighted toward regions north of the equator. That has changed. Today, the remote regions of both northern and southern hemispheres host observatories with large telescopes. Those distant areas have sky conditions which are well-suited to astronomy, well away from light pollution and the other encroachments of civilization. Plus, the southern hemisphere skies feature the centre of the galaxy higher in the sky and there are many other scientifically interesting objects which can only be studied from places south of the equator. For these and many other reasons, professional observatories have spread southward.

The Evolution of the 'Modern' Observatory

State-of-the-art observatories have existed in every age. What changes is technology. From the time of the Renaissance until well into the 20th century, astronomy changed rapidly thanks to the invention of the telescope. The development of more advanced instruments brought the universe closer to the observer. Eventually, by the late 19th century, astronomy began a remarkable transformation as scientists married principles of physics to the study of the stars. That also changed the observatory from a viewing place to a laboratory. How did this transformation happen?

Although there is ample evidence of early observatories built in antiquity (as mentioned in Chapter 2), the first 'modern' observatory has always been considered to be Uraniborg on the island of Hven in Denmark. There, Danish astronomer Tycho Brahe custom-built observatory structures in 1576. It was dedicated to scientific observations of the sky, although it also appeared to have distinct astrological purposes. The building (with its two towers, offices, housing spaces, an alchemical laboratory, and extensive gardens) had no telescope, since the instrument hadn't been invented yet. It did feature a quadrant, which Brahe and his assistants used to measure the altitude of stars and other sky objects as they passed over an imaginary line directly overhead called the 'meridian'. It also had a 1.6-metre (5.2-foot) globe which Brahe used to record stellar positions by inscribing them onto the surface. He also used it to confirm celestial coordinates. The observatory also had additional quadrants, armillary spheres, and a sextant, all for measuring stellar positions.

During his tenure at Uraniborg, Brahe observed what he called *de nova stella* in 1572, which turned out to be a supernova. He also studied the comet of 1577, and determined that its orbital path took it out beyond the Moon. This was a startling conclusion in an age when distances to celestial objects weren't well-understood. Brahe also spent a great deal of time devising a solar system model which had each of its bodies orbiting the Sun in perfect circular paths, an idea astronomer Nicolas Copernicus had proposed a few decades earlier.

Brahe's observatory was supported by the government (as most research facilities in many countries are today) because he talked the king into funding his work. It was also as much a place for Brahe to entertain members of royalty and other important friends, as it was a scientific venue. The place remained open until Brahe lost his funding support and left the country for greener pastures. Unfortunately, after his death, Uraniborg and a smaller, underground facility called Stjerneborg, were torn down. Still, his observatory's influence was felt across Europe during its heyday, when astronomers across the Continent travelled there to study under Brahe and learn his techniques.

Buildings with telescopes inside them really didn't start to appear until well into the 17th century. The Leiden Observatory, for example, arose in 1633 at the University of Leiden. Originally, it was simply a viewing platform created to house a quadrant and other instruments and was located in the centre of campus. In 1860, the university built a new observatory in a botanical garden, and installed a 30-cm (12-inch) telescope along with some smaller instruments. Those instruments were moved to Hartbeespoort in South Africa in 1954. While Leiden maintains an astronomy and astrophysics educational unit, it still refers to itself as an observatory. However, professional observations are no longer conducted there.

Other 'modern' observatories popped up around Europe starting in the mid-1600s. Many were for public study of the stars, but also a number of private facilities began appearing on the estates of wealthy aristocrats, as well as in academic settings. For example, astronomer Johannes Hevelius (who was a civil servant in Danzig, Poland) built the Sternberg Observatory in 1641 as a private facility for his own work. It housed a 46-metre-long (150-foot) Keplerian telescope of his own design. He received many visitors there over the years, including Edmund Halley (of Comet Halley fame), and various members of the court, including the Queen of Poland.

In Denmark, the astronomer Christian Longomontanus, who was also the first professor of astronomy at the University of Copenhagen, suggested the idea of an observatory and received state funding to build it. Brahe's observatory was long gone, and so he built the famous Rundetaarn Observatory on a nearby hill. It was part of a larger set of buildings called the Trinitatis Complex and Longomontanus became the facility's first director. In 1728, a fire swept through the complex and damaged the observatory. It was quickly rebuilt, but was only in use until the early 1800s. Rundetaarn still exists today as a tourist attraction, but astronomy research was moved to a newer observatory in 1861.

The 18th century saw an expansion of observatories to other countries around the world, including new installations in India, Spain, and Russia. In 1774, the Vatican Observatory was established as a centre of astronomical learning. As mentioned earlier, the Church had a long-standing interest in the use of astronomy to determine the high holy days and other events of religious importance. However, it also maintained an interest in the science of astronomy. In the 1800s, some of the first work on stellar classification was done at this observatory, led by the Jesuit priest Father Angelo Secchi. His pioneering research was the first extensive use of spectroscopy to study stars and he used it to make extensive studies of the Sun, which he stated was a star. He devised five classes of stars based on his work at the Vatican Observatory.

At first the facility was located in Rome, but was later moved to various locations. As with most facilities built within cities, by the early 20th century, the Vatican Observatory had to deal with increasing light pollution, smoke, and other challenges to useful observing. In 1961, it relocated to the grounds of the papal residence at Castel Gondolfo, some 25 kilometres (15 miles) outside Rome. Today, the Vatican Observatory Research Group administers the facility, and also has offices and a facility at Steward Observatory in Arizona.

Astronomy in Ireland
Across the Irish Sea from England, astronomers were keen to build observatories as well. Two of the best known 18th-century facilities

are the Armagh Observatory, in Northern Ireland, UK, and the Dunsink Observatory near Dublin, Republic of Ireland. Dunsink was founded in 1785 and the Armagh facility in 1789. The former was funded by an endowment from a former provost of Trinity College Dublin named Francis Andrews. The original funding came to about £3,000 and enabled the purchase of a 30-cm (11-inch) Grubb refractor.

The main work at Dunsink was originally in determining accurate star positions. Among other things, that research allowed astronomers to determine the correct local time standard for Dublin, just as Greenwich did for England. The observatory was in use until the light pollution from the nearby city grew so bright that the facility could no longer be used for precision star measurements. Today, it is part of the School of Cosmic Physics at the Dublin Institute for Advanced Studies and offers public observing nights each season.

The Armagh Observatory was the fulfilment of an interest in astronomy by Richard Robinson, then Archbishop of the Church of Ireland. He was a well-educated, wealthy landowner as well as being the primate of his church and he wanted to build an observatory. His friend Reverend J.A. Hamilton had a small private one and had made observations of the transit of Mercury in 1782. Robinson undertook to build a working observatory for scientific investigation of the skies, and, in particular, for astronomers to use their observations to calculate accurate positions of the stars, and do other scientifically important research. The first telescope installed there was made by Troughton of London, and recommended by Robinson's friend, Neville Maskelyne, the Astronomer Royal of England. In 1834, a Grubb 38-cm (15-inch) reflecting telescope was installed at Armagh, which was used until the 1920s, when it was discarded for parts. Later on, the observatory obtained a 25.4-cm (10-inch) telescope. Its third director, J.L.E Dreyer, used it to study spiral nebulae and other objects he had observed at the private observatory built at Birr Castle by the Earl of Rosse.

In the 20th century, Armagh Observatory joined with the Dunsink and Harvard College in a collaboration on a southern hemisphere instrument called the Armagh-Dunsink-Harvard telescope. It was built by Perkin Elmer in the United States, installed in 1950 at Boyden

Observatory in South Africa, and was the first international telescope in the world. The telescope was used until the 1980s when it was dismantled due to better, larger telescopes being built elsewhere in the southern hemisphere.

The Armagh Observatory facility in Northern Ireland still stands today as part of an Astropark. Some two dozen astronomers work there, carrying out astronomy research. There's also a museum and planetarium nearby, which feature daily presentations and tours.

Beyond Europe

Observatory building also took place in India, with a collection of solar observing and positional astronomy sites, each with the name Jantar Mantar, at eight sites around the country. The best known of these monuments is one at Jaipur in Rajasthan and another in Delhi. Its unique architecture comprises eighteen different astronomical instruments completed in 1734. They were built by the Rajput king, Sawai Jai Singh. The instruments were designed and placed to measure such things as the declination of the Sun at different times of day, the position of the pole star, the longitude and latitude of celestial bodies, measure hour angles, declinations, altitudes, azimuths, and other markers. Some were used to predict eclipses, others to track the location of bright stars, and calculate ephemerides (the tracks of celestial objects in the sky). There were no telescopes installed at the site, although the king apparently did have a private observatory with a telescope of his own. One was built in Madras.

Some other observatories were built in Indonesia during a short-lived 'renaissance' of scientific research done there by European settlers. The Mohr Observatory was built by the Dutch-German pastor John Marits Mohr. (Indonesia was known as the Dutch East Indies in his time.) He had offered his estate near Batavia (now Jakarta) for observations during the 1761 transit of Venus. Mohr was an avid astronomer, and after the transit, he decided to build a private observatory outfitted with appropriate instruments such as a meridian, a parallax 'machine', wind meters, a compass and a reflecting telescope. All were delivered in time for him (and invited scientists) to view the next transit of Venus and a transit of Mercury in 1769.

The Royal Observatory at Greenwich

The history of the Royal Observatory is a long and honourable one. It became renowned for its timekeeping service, supplying Greenwich Mean Time, and also for the meridian which runs through the property. That line splits the Eastern Hemisphere from the Western Hemisphere, and remains a favourite of visitors. While we commonly associate astronomical observatories today with telescopes and other instruments, many early ones, such as the Royal Observatory, were built originally to advance navigation and timekeeping techniques.

Seafarers and other travellers looked to the stars for guidance as they moved around our planet. In particular, knowing one's position on the planet required accurate positions of the fixed stars, which would help give a traveller's latitude. For example, the North Star (called Polaris), helps people know where north is. By measuring its height above the horizon, to know how far north or south one is, particularly in the northern hemisphere and parts of the southern hemisphere. Other stars can also be used for latitude determination. The question of longitude, how far east or west one's position is, remained quite a challenge until well into the 17th century. Thus, the Royal Observatory was created for observations of the sky that would use time and the positions of the stars to help sailors and others determine their positions at sea.

Charles II granted the funding for the observatory in 1675, and it was built at Greenwich, which lies along the Thames River about 9 km (5.5 miles) from London. The oldest building on site is called Flamsteed House, named for astronomer John Flamsteed who laid the cornerstone for the observatory as its first director. He was appointed the first Astronomer Royal in England in 1675, and spent 44 years observing the sky and charting stellar positions. From this work, he developed a positional atlas called *Atlas Coelestis*, which was updated and published after his death.

The observations done at Greenwich were largely concerned with what's called 'positional astronomy', that is, the accurate positions of stars. The Greenwich astronomers used telescopes designed to look at the sky on a north-south axis, essentially along a meridian line. The line which crosses through Greenwich on the grounds of the observatory is called the 'Prime Meridian' and demarcates

0 hours of longitude. As stars move across the sky (and cross the meridian), their time of crossing can be tracked and noted. They appear to move east to west, just as the Sun does. Charting the crossings each day to get an accurate sidereal time was done at Greenwich, making this observatory one of the world's 'timekeepers' by default. The term Greenwich Mean Time (GMT) stems from the observatory's primary mission of determining accurate time through stellar observations.

Over the years, the Astronomers Royal oversaw improvements in the timekeeping process, and eventually, over the publication of a Nautical Almanac. That well-known book included tables for stellar positions, and made seagoing navigation and timekeeping less of a challenge.

As the science world's attention shifted from purely observational astronomy to astrophysics, the Royal Observatory constructed a state-of-the-art 71-cm (28-inch) refractor which came be called the Great Equatorial Telescope. It was commissioned in 1885 and completed in 1893, and is well-known for its observations of double stars. It was also used for astrophotography up to the time of its decommissioning in the late 1960s. Rather than lose any of the telescope's capabilities, astronomers refurbished it with computers and CCD cameras and returned it to use for educational programmes. It continues to be used to this day.

Up until the 19th century, the main business of astronomy was focused on measuring the positions and computing the motions of celestial objects. By the mid-1800s, a new science, called 'astrophysics' was coming into play, as astronomers began looking first at the spectrum of the Sun and then spectra of the stars. This 'spectroscopic' method swiftly transformed astronomy.

Even with the rise of this 'new' science of astrophysics, the Royal Observatory's main mission remained positional astronomy for quite some time, using the best telescopes it could afford. Over time, it became clear the site at Greenwich wasn't the best for scientific operations. Light and atmospheric pollution were threatening the view of the skies. The observatory's magnetic observatory was also in need of a new location.

After the Second World War, the Royal Observatory's programmes were sent their separate ways. Most functions for astronomy went to Herstmonceux (a village in Sussex), although delays in moving the telescopes meant that their technology was rapidly surpassed by newer instruments. A new telescope, the Isaac Newton, was built and installed at the new site, but then moved to Cambridge. Other observatory functions and staff moved to the Royal Observatory Edinburgh, some remained at Herstmonceux, and other functions (including the Nautical Almanac offices) moved to Rutherford Appleton Laboratory in Oxfordshire. The old Royal Observatory closed its doors in October 1998 after a remarkable 323-year record of operations.

In recent times, the original site at Greenwich has been taken over by the National Maritime Museum, which maintains exhibitions of telescopes, nautical instruments, the sailing ship *Cutty Sark* and the Prime Meridian. The Peter Harrison Planetarium is also on site.

Sydney's Observatory

In Australia, Sydney's famous observatory that visitors can see as they walk through Circular Quay or spot from the nearby Harbour Bridge, was built in 1858. Like its sibling observatories in Ireland and England, it was primarily planned for timekeeping purposes, star chart making, and astronomy research. The building was equipped with a time ball, which dropped each day at 1 p.m. The current site is called Observatory Hill. For much of its history, Sydney Observatory was involved with creating almanacs for navigation, and played a role in meteorology and timekeeping. It was also a pivotal site for observing the southern sky.

Today, the facility is surrounded by the city, and – due to light pollution – is no longer a research institution. It's now part of the Museum of Applied Arts and Sciences, and offers exhibits about the observatory's history, plus stargazing on observatory nights, solar viewing, astronomy courses, and shows in its planetarium. It contains the oldest working telescope in Australia, plus a modern computer-controlled instrument, and a solar telescope.

Changing Location to Improve the View

For a long time, people such as Herschel and Messier catalogued objects, and carefully tracked the positions of stars and planets. It was only a matter of time before observers began to look beyond accurate positions and solar measurements to search for an understanding of what those objects actually are. This interest influenced the growth of astronomy as a branch of science. For that, new tools and practices were needed.

The scientific application of maths and physics to astronomy required better telescopes inside well-equipped observatories. The operative word here is 'inside' because putting a telescope inside a building, or some other kind of enclosure, protects it and the instruments attached to it from the elements. And, with the advent of larger, more complex telescopes, protecting them from the extremes of weather became more important.

Just as our earliest ancestors did with some of their observing spots, it didn't take for astronomers to realise observatories needed to be located in remote sites. There are several reasons for this. Our atmosphere plays a huge role in determining the 'seeing' that optical (and infrared) telescopes can do from the ground. We can get a good idea of what astronomers mean by 'seeing' by simply stepping outside on a clear dark night. Even on the clearest, quietest night, a casual observer can see the stars twinkle and planets may appear to 'waver'. This is due to the movement and mixing of masses of air in our atmosphere. Not only does it perturb the view that ground-based instruments get, but our atmosphere also absorbs some incoming radiation. It's also why some telescopes are located in space, well above the blanket of air.

Astronomers can't send all their telescopes to space, however. There are places on the planet where seeing is quite good. The best conditions exist high atop mountains, at high-altitude observatories. That's where atmospheric distortion is much less of a problem than it is at lower regions. In addition, observatories wanting to do observations in the infrared range of the spectrum have much better conditions in higher, drier climates. That's because infrared is absorbed by water vapour and atmospheric carbon dioxide. Typically, the air higher up has less of these. Astronomers as far back as Charles Piazzi Smith (appointed Astronomer Royal for Scotland in 1846) knew from their travels around the world for eclipses and other celestial events that high, dry climates were better for making crucial observations.

The Great Refractor of the Paris Observatory at Meudon. It is an 83-centimeter (32-inch) refractor that was built in the 1870s. Courtesy of Observatoire de Paris.

Atmospheric aberration isn't the only problem high-altitude observatories were built to avoid. Light pollution is a consistent problem at lower altitudes. Since the age of electric lights began in the 19th century, stray light has plagued astronomers. So, they began building where it was high, dry, and dark.

One of the largest telescopes of the 19th century was installed outside Paris at Meudon, and was known as the Meudon Great Refractor or 'Grande Lunette'. Astronomers used it to study the planets, comets, double stars, and asteroids. For many years, it was among the most sought-after telescopes in Europe. The Paris Meudon Observatory also has a long and honourable history. It was initially established as the Paris Observatory during the reign of Louis XIV. He was convinced to fund such a place to help astronomers in their work of observing and predicting the motions of celestial bodies. This, in turn, would improve information for navigators. Astronomers had grand plans for the observatory – it would provide laboratories, workshops, a chemical laboratory, and a place for the newly founded Academy of Sciences (1666) to hold its meetings. The building was ready for use in the 1670s, and the famous Jean Dominique Cassini served as an academic advisor. His observations of the Moon and Saturn (among other objects) followed. The Paris Observatory merged with the Meudon facility in 1926, already part of the new breed of facilities that were combining telescopes and laboratory work, using spectroscopes (and eventually cameras) to study celestial objects. Over the years, it also incorporated a coelostat, a heliograph, and other instruments. The observatory has been expanded several times, and eventually came to be the headquarters of the International Time Bureau, which provides standard time determinations for other observatories. Today, the observatory (which is administered by Université PSL in Paris), has a solar telescope, a radio telescope at Nançay, and maintains a collection of books, data, and observing archives related to the observatory.

Another good example of the 19th-century desire to 'move up and out' lies in the south of France. High in the Pyrenees, a group of scientists began construction of a high-altitude facility, Observatoire du Pic du Midi, in 1878. It was a costly venture, and eventually the French government took over the site and saw it through to completion. This was an observatory that truly bridged the centuries, and delays enabled its builders and director, Benjamin Baillaud, to installed state-of-the-art equipment. In 1908, the observatory dome was completed, which housed a reflecting telescope. The observatory also acquired a 0.60-metre (23.6-inch) telescope in 1946, and a spectrograph was installed in 1958. In 1963, Pic du Midi Observatory participated in the US effort to send missions to the Moon by supplying detailed lunar images with a 1.06-metre (42-inch) telescope funded by NASA. Today, the observatory has the 2-metre (6.5-foot) Bernard Lyot Telescope (which is the largest in France), plus a coronagraph for solar studies, and a 0.60-metre instrument used mainly by amateur observers. Among other discoveries, the Pic du Midi instruments have been used for exoplanet discoveries, studies of asteroids and solar system moons, and was among the first observatories to discredit the so-called discovery of canals on Mars.

In Imperial Russia, the Academy of Sciences built the Pulkovo Astronomical Observatory some 20 kilometres (12.5 miles) south of St Petersburg in a region called Pulkovo Heights. It was only about 250 metres (820 feet) above sea level, but far enough away from the city to get good views. The observatory opened in 1839, headed by astronomer Friedrich Georg Wilhelm von Struve. It contained a 38-cm (15-inch) refractor, and later on a 76-cm (30-inch) Alvan Clark refractor was installed. These were among the largest in the world at the time. While astronomy was one of the main purposes of the observatory, it was also used to make star catalogues, particularly of the polar regions of the sky. It was also a main timekeeping centre, and in 1920 began transmitted time signals throughout Russia. Pulkovo was damaged heavily during the Second World War, although some of its mirrors and library were saved from the encroaching German forces. After the war, the Soviet government restored the facilities for research, including a new radio astronomy division, and a facility for instrument making. A 66-cm (26-inch) refractor was installed, a stellar interferometer, new solar telescopes, a coronagraph and other instruments were added to the observatory's toolkit.

The United States Naval Observatory, established in 1830 in Washington DC, primarily as part of the country's Depot of Charts and Instruments. Its main mission was to handle the navy's charts, navigational equipment, and chronometers. In the mid-1800s, the observatory began publishing astronomical almanacs and nautical publications. These almanacs provided the most precise stellar positions for both observers and sailors. The observatory also became a centre of timekeeping and used a time ball that dropped down a mast to indicate mean solar noon for navigators aboard ships anchored outside the city. The observatory eventually expanded its mission to a wider variety of scientific projects, such as the study of solar eclipses, and sent astronomy expeditions to view the transit of Venus. USNO astronomers also worked to measure the speed of light. In 1870, the US government granted the observatory enough funds to buy and install a 66-cm (26-inch) Alvan Clark telescope. It was used mainly to study planetary motions, and was also the instrument that USNO astronomer Asaph Hall used to discover the moons of Mars, Phobos and Deimos. Like many other cities of its era, Washington DC was growing. Soon, the town had surrounded the observatory, and its location – called Foggy Bottom – wasn't the best location for a precision telescope. Not only was it next to a river in a humid climate, but astronomers claimed it wasn't a healthy place to work. So, the observatory moved to its current location, at the time a more rural site north of Georgetown. There, the conditions for seeing were better, and the observatory continued its work in timekeeping, while adding new programmes for solar monitoring. The observatory remains an important source of timekeeping and celestial observations, and continues to publish its almanacs for observers and navigational needs. Today, of course, the Naval Observatory is surrounded by the greater Washington D.C. metropolitan area, but it remains open to visitors. Its grounds also contain the official home of the Vice President of the United States.

Spectroscopy and Astronomy

Today, we completely understand the placement of observatories where they can get clear 'seeing'. While great views are a main driver, there's another reason: spectroscopy. This scientific discipline really began with Isaac Newton's work in optics and light and has been a staple of chemistry (where scientists measure the different spectral lines of the elements). In the 1800s, people such as German chemist Joseph von Fraunhofer began doing experiments to separate light into its component wavelengths. He used prisms (which actually fascinated the early Romans), and then later instruments called spectrometers.

In 1859, Gustav Kirchhoff and Robert Bunsen (inventor of the Bunsen burner every chemistry student is familiar with) reported something curious. As Kirchhoff heated certain chemical elements in the burner, and as he viewed the resulting light through a spectral instrument, bright and dark lines appeared in the spectrum. These were from the chemical elements being burned, and each one had a specific 'fingerprint' of bright and dark lines. Kirchhoff went on to study and identify at least sixteen different chemical elements by their spectra, and associated those with spectra he took of the Sun. It was a pivotal moment that represented the marriage of chemistry and astronomy.

For a time, astronomers concentrated on solar spectra, since our star is bright and easy to observe. Eventually, they began to apply the new science to the study of stars. From this work, the concept of stellar classification was born. Why classify stars? Just a casual glance at the sky on a dark night shows that stars have different hues. Some, like Betelgeuse or Aldebaran, appear reddish-orange. Others, such as Sirius, appear blue-white. Our Sun is yellow-white. Classification by colour is easy, but what do those colours mean?

In the beginning, astronomers used three classes: blue/white stars, yellow (solar-type) ones, and red stars. This all changed when Harvard's Edward C. Pickering began a huge stellar classification programme using spectra recorded on photographic plates – a process first used by astronomer Henry Draper. Pickering began the project, aided by his female 'computer' Annie Jump Cannon, in 1885. By 1990 they had classified a catalogue of more than 10,000 stars. Eventually, the catalogue expanded to nine volumes, and by 1924, there were ten spectral types that students still learn today: O, B, A, F, G, K, M, R, N, S. They represent a rough evolutionary guide to stars. O and B stars are hot, young (and often massive) stars. G stars, like our Sun, are roughly middle-aged. Those classified as M and beyond generally tend to be older and cooler. Of course, the devil is in the details, and there are other characteristics which help astronomers 'date' and 'classify' stars, but this early work led the way to more detailed classifications in the 20th century.

Today, astronomical spectroscopy is used to study motions and characteristics of star clusters, nebulae, galaxies and galaxy clusters. The spectrum of an object reveals information about its temperature, mass, motions, brightness, magnetic fields, and chemical composition. And astronomical spectroscopy, like most other forms of observation, depends on clear, steady seeing.

Photography and Imaging in Astronomy

The 19th century continued to see more such observatory expansion across the world, including sites in South America. In addition, after the Civil War in the United States, many universities began building their own observatories. Apart from spectroscopy, many facilities also took advantage of another new invention: photography. The application of photography to astronomy in a scientific way really was part of the transformation of astronomy from a purely observational pastime to one of scientific inquiry. Observers began using early cameras to make long-exposure photographs of the sky. Lengthy exposures required the telescope to remain still, otherwise the image would be out of focus. Prior to the 19th century, observers simply moved the telescope to aim at different parts of the sky and would periodically 'nudge' it along to keep the object in the field of view. Photography and (later on) spectroscopy required a very steady view of the sky. The advent of those two practices led to the need for a way to keep the telescope 'on track' throughout a long observing session. Telescope makers built clock drives to move the instrument at a constant rate while it was focused on an object. They also had to improve the mounts to keep the telescope aimed accurately without moving 'up and down' and 'side to side'. There were two main types of telescope mounts developed over the years that would allow such movement: the altazimuth mount and the equatorial mount. Both allow the telescope to be pointed as needed. A clock drive held it in position during an observation.

With advances in mounts and tracking, astronomers could take images of the sky and spectra of specific objects for further, later study. The first astrophotography attempts were made by Louis Jacques Mande Daguerre. We know him better as the inventor of the daguerreotype, an early photographic process. He took the first image of the Moon in 1839, and astronomers saw great promise in his work, although it took several decades before astrophotography was widely used. By the late 19th century, astronomers were using dry plates for both imaging and spectra. The 20th century saw the rise of films for astrophotography, and by the end of the century, CCD cameras were part of the digital revolution that overtook astronomy.

The Age of the First Great Telescopes
The first of the great telescopes aimed solely at astronomy research appeared on the scene in the mid-to-late 19th century. In addition to those examined above, there were:

Dearborn Observatory (Chicago, IL, 1862)
Imperial Observatory (Strasbourg, France, 1880)
Royal Observatory (Edinburgh, Scotland, 1872)
Quito Astronomical Observatory (Ecuador, 1875)

The National Observatory of Paris (1875)
Imperial Observatory (Strasbourg, France, 1880)
McCormick Observatory (Charlottesville, VA, 1883)
Chamberlin Observatory (Denver, CO, USA, 1891)
Lowell Observatory (Flagstaff, AZ, USA, 1896) and many others.

As each facility was built, improvements in telescope design and instrumentation were part of a continuing evolution in observatories – turning simple observing buildings and platforms into places where more complex science studies could be done.

The Rise of the Observatory as Laboratory

As observatories began to dot the global landscapes, astronomers started to develop them as more than simply places to observe the sky. They wanted science labs. We have already read about the rise of the observatory as a laboratory with the Paris Meudon facility in France and the Royal Observatory at Edinburgh in Scotland. In the United States, Yerkes Observatory, built in Williams Bay, Wisconsin, was the first one to really be purpose-built as a research facility and laboratory. It was founded by American astronomer George Ellery Hale and has been operated by the University of Chicago Department of Astronomy and Astrophysics.

Hale was an interesting fellow. He fell in love with astronomy as a young man, when his father gifted him with a telescope and a place to use it called the Kenwood Astrophysical Observatory. With his 30-cm (12-inch) refractor, and a spectroheliograph he invented while a student at MIT, at the observatory Hale indulged in his passion for solar studies and photography. He was the first to study outbursts and solar prominences. His work expanded the use of spectroscopy, and when he moved on to build his first telescopes, his earlier instruments were donated to Yerkes. His passion for the Sun carried over to his establishment of other observatories.

At Yerkes, astronomers were expected to carry out scientific experiments. This was based on Hale's insistence that an observatory should be a science laboratory, albeit for objects in the sky. He wanted to apply physics to star studies (astrophysics) and spectroscopy (chemistry) to understand the characteristics of objects. Under Hale's

The 1-metre (40-inch) telescope installed at Yerkes Observatory in Williams Bay, Wisconsin. Gathered in front of it are physicist Albert Einstein and the staff of the Observatory, with astronomer E.E. Barnard, back row. Courtesy of Yerkes Observatory.

direction, plus that of his successors, Yerkes became a training ground for the next generations of astronomers.

During its time, Yerkes was used by some very well-known astronomers, among them Subrahmanyan Chandrasekhar, Edwin Hubble, Gerard Kuiper, Albert Michelson, Carl Sagan, and Otto Struve. The observatory remained a student training facility up until its 2018 shuttering, and also welcomed science educators and the public to workshops and other programmes.

Astronomy at Yerkes

Yerkes started its 'life' equipped with a 102-cm (40-inch) refractor containing Alvan Clark lenses and has been used over the years to create nearly 200,000 photographic plates of the sky. It also houses two reflecting telescopes, a 100-cm (40-inch) and a 61-cm (24-inch) instrument, plus smaller telescopes. Throughout the years, astronomers have made major discoveries at Yerkes in fields such as Milky Way studies, stellar spectroscopy, planetary atmospheres, the mechanics of interstellar dust clouds, cometary studies, the physics of white dwarfs, and stellar distances. The observatory operated for more than a century as both a professional observatory and training centre until the university closed it down in 2018 and offered it for sale. Part of the problem was with the 'seeing' at the observatory. Light pollution from the nearby cities of Milwaukee and Chicago had degraded the sky view. In addition, upkeep of the building was starting to become costly. The university made the difficult decision to divest itself of the property, and the future of the building and its instruments is uncertain.

Observatory Boom Times

The late 19th century saw a boom in the construction of other research observatories, each one larger and more complex than the last. European facilities, such as the Astrophysical Observatory of Potsdam, were built specifically to advance the physics of the stars and planets, and it was there that a great many physics-related experiments were performed. For example, scientists at Potsdam studied radial velocities of stars, searched for solar radio emissions, and eventually played host to the invention of photoelectric photometry (the measurement of brightness of celestial objects). By the middle of the 1800s, Potsdam also saw developments of what would become the science of spectroscopy. Gustav Kirchhoff and Robert Bunsen jointly devised the science of spectral analysis, which allowed astronomers to get the chemical 'make-up' of stars by analysing their light. It also led to the creation of a solar observatory in 1871. The formal founding of the Astrophysical Observatory of Potsdam took place in 1874, and the main focus of the work was stellar astrophysics.

By the century's end, Potsdam could boast what was then the largest refractor in the world, with 80-cm (31-inch) and 50-cm (20-inch) lenses. The telescope was mounted under a 24-metre (78-foot) dome and dedicated to science by Emperor Wilhelm II. Use of this refractor heralded a number of great discoveries, including the detection of various emission lines in the Sun and other stars. Researchers such as Karl Schwarzschild (best known for his work on Einstein's calculations relating general relativity to such objects as black holes) made fundamental contributions to astrophysics using this facility. Today, the Potsdam Observatory is under the administration of the Leibniz-Institute for Astrophysics Potsdam, having survived relocation, Second World War damage, and the reunification of the two German States in the 1990s.

Astronomy in the US

The United States continued to see an expansion of observatory building at universities. The Harvard College Observatory began operation in the late 1700s. Through the efforts of William Crunch Bond, its first official 'astronomical observer' the facility acquired a 38-cm (15-inch) 'Great Refractor' in the 1840s. That acquisition helped establish the observatory among the earliest of the astronomy research institutions. The great astronomer William C. Pickering furthered its development later in the century, when he established a programme of both visual observations and photography, as well as the application of stellar spectroscopy. The observatory still maintains its famous plate collection, worked on and analysed by the likes of Annie Jump Cannon, Henrietta Swan Leavitt and others.

In the 20th century, the Harvard College Observatory continued its research programmes under the leadership of Harlow Shapley. He created a graduate programme in astronomy and astrophysics, training such scholars as Cecelia Payne-Gaposchkin, Helen Sawyer Hogg, Carl Seyfert (of Seyfert galaxies fame), Dorrit Hoffleit (who went on to create the Yale Bright Star Catalog), and many

The Harvard College Observatory in Cambridge, Massachusetts. Its first telescope, the 38-cm (15-inch) 'Great Refractor', installed in 1847, was the largest telescope of its kind for twenty years. It was one of the first telescopes to be used for astrophotography, producing the first images of the Moon. Public domain image.

others. Some of these students made fundamental discoveries in astrophysics, notably Payne-Gaposchkin. Her doctoral dissertation focused on stellar atmospheres and, in particular, she concluded that the composition of the Sun was primarily hydrogen. It was a notable piece of work, but her advisor (Henry Norris Russell), didn't think it was correct. Later on, he changed his mind and took credit for her discovery. Today, she is known as the woman who discovered what stars are made of, and her work on variable stars laid the basis for further study. She later became chair of the department of astronomy, and she supervised a number of students who went on to make ground-breaking discoveries in astrophysics throughout the 20th century.

The Harvard College Observatory came under the purview of the Harvard department of astronomy in 1931, and Donald Menzel became its director in 1952. The Smithsonian Astrophysical Survey relocated to Cambridge in 1955. Both departments were later united under the umbrella of the Harvard-Smithsonian Center for Astrophysics. The Harvard Observatory remains on the campus at Harvard, and its scientists, researchers, and students can use an array of ground-based observatories in Arizona, Hawaii, and Chile, as well as having access to space-based telescopes such as Hubble Space Telescope, Spitzer Space Telescope, and the Chandra X-ray Observatory.

Beyond Harvard
The Hopkins Observatory at Williams College in Williamstown, Massachusetts, is another typical college facility. It was originally built in 1838, and the college claims it is the oldest one in the continental United States. It has moved several times during its history, and the original building now contains a planetarium and several of the original instruments. Currently the observatory, under the directorship of Dr Jay Pasachoff, has a 60-cm (24-inch) Cassegrain telescope, a spectrograph, and several smaller refractors and reflectors, plus a solar telescope. The instruments are largely for use by the students, and the college has an active observing programme for solar eclipses.

Another college facility, the Dearborn Observatory, was built in 1888, and joined Harvard and Pulkovo Observatories in having telescopes 38 centimetres (15 inches) or larger. It was initially located in Chicago and featured an Alvan Clark refractor that was originally built for the University of Mississippi, but went unused. The Chicago Astronomical Society and University of Chicago acquired the lens and used it as the heart of a new observatory. In 1887, the telescope was moved to Northwestern University, and has been used ever since for a variety of studies and for training in the university's astronomy department.

Go West, Young Astronomers!

To accomplish a 'modern age' of astrophysics and improved astronomy, the people who wanted to build new observatories looked for better observing conditions in drier climates. In the US, the western half of the country beckoned to astronomers. Over the next few decades, mountaintops in Arizona and California sprouted observatories that made significant contributions to astronomy.

Conditions were especially enticing in California, and drew a generation of astronomers to locate telescopes throughout the state. The late 19th and early 20th centuries comprised an age of wealthy men interested in patronising these new institutions, as much for their love of astronomy as for the fact they wanted to leave behind showy educational and scientific legacies. Philanthropists, real estate developers, steel magnates, and others donated generously to this new breed of institutions.

The Lick Observatory, which bills itself as the first permanently occupied mountaintop facility, is located on Mt. Hamilton, near San Jose, California. It's named after James Lick, who donated a record (for the time) $700,000 to build the facility and create a good road to get to the observatory. Construction began in 1876, with wagon trains of materials being dragged to the top of the mountain. The first telescope installed there was a 30-cm (12-inch) Alvan Clark refractor. It saw its 'first light' in 1887 and such observers as E.E. Barnard did their early work there to photograph comets and nebulae. The observatory's second telescope was a 91-cm (36-inch) refractor which came into use in 1888. For several years, it was the largest refractor in the world – until the installation of the Yerkes telescope a decade later.

The observatory was assimilated into the University of California system in 1888. Today it serves astronomers throughout the university system, and its users range from research fellows to undergraduate students. Among the many discoveries made using Lick Observatory have been the moons Amalthea, Elara, Himalia and Sinope orbiting Jupiter, in 1905. The observatory has also been used to measure the sizes of Jupiter's largest moons, conducted extensive asteroid observations, and – more recently – has made discoveries of a number of exoplanets. Astronomers have also used it to take spectra of distant galaxies, study the evolution of stars and galaxies, and do continuous monitoring of active galactic nuclei. In addition, Lick Observatory astronomers and technicians worked with scientists at Lawrence Livermore Labs to create the world's first laser guide star system for adaptive optics. This laser-beam-based system allows astronomers to measure and correct for atmospheric conditions. The data it produces helps produce sharper images of celestial objects.

Today, Lick Observatory is equipped with the 3-metre (120-inch) C. Donald Shane reflector, the Anna L. Nickel 1-metre (39-inch) telescope, the original Lick 91-cm (36-inch) instrument, the Crossley 90-cm (35-inch) telescope, plus assorted spectrometers, spectrographs, cameras, and its specialised adaptive optics system. Lick is also the home of the Automated Planet Finder 2.4-metre (94.5-inch) reflector, as well as several smaller telescopes. The observatory also serves as a testbed for new equipment, and welcomes members of the public up the mountain for various observing programmes.

Astronomy in the High Desert

Among the astronomers looking west for favourable viewing places was Percival Lowell. He was from a wealthy Boston family and had spent his early years working in business and traveling in the Far East before following his passion for astronomy. In particular, Lowell was interested in Mars and wanted to observe it as much as possible. To do that, he searched for a place with good observing conditions. He chose the high desert location of Flagstaff, Arizona, at a site which eventually became known as 'Mars Hill'. Lowell equipped the observatory with a 61-cm (24-inch) Alvan Clark telescope. It was what he used to study the Red Planet for many years. Most people also know of him as the man who wrote extensively about the possibilities of life on Mars, and his search for the so-called 'canals' on the Red Planet.

Lowell Observatory is also famous for Clyde Tombaugh's discovery of Pluto, using a 33-cm (13-inch) telescope. That instrument is now known, appropriately enough, as the Pluto Discovery Telescope and it underwent renovation and restoration in 2014 as one of the observatory's historical artefacts.

Pluto's discovery wasn't on a whim. In addition to looking for canals on Mars, Percival Lowell was fascinated by the idea there might be a Planet X 'out there' in the far reaches of the solar system. He looked for it from 1905 until his death in 1916, but never was able to find any trace of a planet. More than a decade after he died, the observatory hired the young Clyde Tombaugh to resume the search using a telescope built for the purpose. It took Tombaugh nearly a year of taking plates of the sky and then carefully examining them. He had to use what's called a 'blink comparator', which held two plates and allowed him to 'blink' back and for the between the two to watch for something that was moving. On 18 February 1930, Tombaugh found Pluto. It looked like a tiny dot on each plate, and it had clearly changed position between the two exposures. Those plates are still on

Percival Lowell observing Venus through the Clark Refractor in 1914. Public Domain image.

display at Lowell Observatory, a testament to the painstaking work he did to locate this distant dwarf planet.

Today, the observatory also has the Discovery Channel Telescope, a 4.3-metre (14-foot) instrument which came online in 2015. It's the 'flagship' instrument, operated by a partnership between Lowell, Boston University, the University of Maryland, the University of Toledo, and nearby Northern Arizona University. In addition to research, the observatory has also hosted a number of special researchers, including Apollo mission astronauts, who trained there and in nearby Meteor Crater for their moon landings.

Current Research at Lowell Observatory

Astronomy research is alive and well at Lowell, with studies in small solar system bodies, exoplanets, stellar evolution, and extragalactic star formation. The observatory has research telescopes on nearby Anderson Mesa. These include the Navy Precision Optical Interferometer, and the 1.8-metre (72-inch) Perkins Telescope, as well as the 1.07-metre (42-inch) John S. Hall Telescope. The observatory is open to visiting scholars, graduate students, and others, in addition to its own research staff. There's a major public outreach programme that focuses on Native American astronomy, solar observing as well after-hours observations through the Clark Telescope.

An Observatory for Denver

Arizona was not the only western state to gain an observatory in the latter years of the 19th century. Chamberlain Observatory in Denver, Colorado, was built by Denver University in 1894, and equipped with a 0.508-meter (20-inch) refractor with a lens made by Alvan Clark and a mount engineered by George N. Saegmuller. As with other observatories of the era, Chamberlain was made possible by donations from a wealthy magnate, Humphrey B. Chamberlain. He made his fortune in real estate and pledged $50,000 to build and equip the observatory. The facility was used for many years by the university's students and faculty. Now totally surrounded by the city of Denver (and its light pollution), Chamberlain is no longer used for research, and is now operated by the Denver Astronomical Society, which holds its meetings there and offers public outreach and stargazing events.

California's Observatories as Bridges to the 20th Century

The turn of the 19th century continued astronomy's growth, not just as a major industry for telescope makers, but for the advancement of astronomy as a whole. Facilities built in the 19th century continued opening up marvellous views of the sky. The next generation of observatories began to rise. The first of these was on Mount Wilson in the San Gabriel Mountains north of Los Angeles, California. It was built to take advantage of good, dry weather conditions and dark skies that were preferable for observations, photography, and spectroscopy. Mount Wilson Observatory (now part of the Observatories of the Carnegie Institution for Science, also known as the Carnegie Institution of Washington) was the brainchild of George Ellery Hale, of Yerkes fame. He worked tirelessly to get more than $500,000 in funding, tapping the Carnegie Institution of Washington to build his brainchild.

The first project on the mountain was the Snow Solar Telescope, and he imported a group of researchers and engineers to work on it. Hale always wanted this facility to be dedicated to solar studies. For a time, it was known as the Mount Wilson Solar Observatory. However, the location was so good that it made sense to put a regular telescope up there. The 1.5-metre (60-inch) telescope was built, followed by the construction of what was then the largest one in the world: the 2.5-metre (100-inch) Hooker telescope.

A local business friend of Hale's, John D. Hooker, after whom the telescope is named, pledged $45,000 to make the large mirror that Hale insisted on having, with some funds left over for the building and ancillary equipment. The glass itself was created by the French Plate Glass Companies of Saint-Gobain, in France, and the 9,000-pound (4,082 kg) slab was made of wine bottle glass. When the time came to grind glass for the Hooker telescope, there were many questions about how well it would work. This was due largely to the size of the telescope Hale had in mind. At first, no one was quite sure if such a large mirror could actually be made. And, if it was cast, would it be stable and useful? That was the challenge facing Hale and the technical team.

Despite its stability, even the best optical glass can contract and expand in response to temperature changes. In addition, even though the weather at Mount Wilson was far more stable than

Hale's previous posting in Wisconsin, the team members thought they might need to find ways to keep the observatory dome at the proper temperature to stabilise the viewing and protect the mirror from huge temperature swings, moisture, and other environmental hazards. In addition, the behemoth mirror would need a very secure tube and a stable mount to ensure good observations. This was, in large part, due to the fact that California is subject to earthquakes. So, the building and mount were made as quake-proof as possible.

Work went ahead and the mirror was cast. While that was happening, the observatory built a place for grinding the mirror, the Hooker Building at headquarters in Pasadena. It had rooms for grinding and figuring the mirror, plus a grinding machine and a 100-foot-long testing hall. Like the observatory, the building was fireproofed and stabilised in case of another earthquake.

The mirror blank arrived in Pasadena in late 1908, but much to the astronomers' dismay, it ended up with bubbles in between layers of the glass and they feared it would be useless. But, after some testing and a lot of heated discussion between optician George Ritchey, John Hooker, and other team members, they went ahead with grinding. What followed was four years of shaping the world's largest telescope mirror, as well as making an appropriate mount, clock drive, and building for it once the glass was ready to be placed into the telescope. On 1 July 1917, the completed mirror was loaded onto a truck and began its final trip up the mountain. Five months later, on the night of 1 November, the Hooker telescope had its first light. Its first view, of Jupiter, was not perfect, but after a few hours of cooling the telescope then showed a perfect single image of a star. It had been a long, journey, but at last, the world's largest telescope was ready for work.

Advances in Glass

The Hooker telescope incorporated many advances in glass-making and mirror grinding. As astronomers were coming to see, glass mirrors were an improvement over the old metal reflectors. Glass holds a focus much longer and better, and once a mirror is 'poured', its general shape is set. Of course, it needs to be ground to a precise shape that will direct the reflected light to a focal point. For a reflecting telescope, a coating is then needed. For refractors, the final pieces of glass have to be ground precisely to focus light properly through them and to avoid aberration.

In 1864, when French scientist Leon Foucault built his first large reflecting telescope using glass, the stage was set for the development of larger, more accurate telescope mirrors made from blanks that needed to be shaped and ground. Mirror blanks are simply that – a piece of glass cast for astronomy use. The blank has to be able to withstand changes in temperature conditions without cracking, and hold its shape as its mounted in a telescope. The bigger the blank, the heavier it is, and the more stress it will need to withstand. While there are a number of single-piece mirrors in use today, they are thinner than their earlier counterparts. In addition, a growing number of facilities are using mosaics made up of smaller flexible mirrors which weigh much less than a large mirror, and can still deliver a reflecting surface for precision astronomy observations.

Today, several companies around the world make glass blanks specifically for telescopes. For example, Corning Glass in New York has created the basic mirrors for Hubble Space Telescope, Subaru, and Gemini Observatories. United Lens Company does the same, and its mirrors are found in a wide range of instruments.

For reflectors, the 'reflective' part is the coating that is applied after the mirror is ground to the correct shape. As mentioned earlier, thin layers of silver work well, as do gold and silver. The glass itself can be made of fused silica, quartz, ceramic, as well as a type of glass called Zerodur. It's a lithium aluminium silicon oxide glass ceramic.

Hubble Discovers the Universe

Mount Wilson Observatory has been home to many spectacular discoveries in astronomy and astrophysics. It bills itself as the place 'where we discovered our place in the universe'. During the first fifty years of the 20th century, the institution became famous world-wide as a place where humans first began to understand about galaxies and the larger universe. The startling discoveries made at Mount Wilson revolved around astronomers' perceptions of our own galaxy, and—eventually—of nearby galaxies. Astronomer Harlow Shapley determined that our Milky Way Galaxy was much larger than astronomers had previously expected. Using the Hooker telescope, he made many observations to measure its size. He thought our galaxy was the sum total of the universe. His idea put him in direct conflict with fellow astronomer Edwin P. Hubble, who made observations at Mount Wilson that showed our galaxy was only one of many.

In Chapter 1, we saw how Hubble made a simple observation of a variable star in a distant 'spiral nebula' (called the Andromeda Nebula), which led to a distance calculation putting that star well outside our own galaxy. By definition then, the Andromeda Nebula was not just a cloud of gas and dust inside our own galaxy, but a galaxy in its own right. It also implied the universe was larger and populated with countless other galaxies. Moreover, further measurements and observations by Hubble and others showed the universe is actually expanding. Those discoveries, made at Mount Wilson, rocked the foundations of astronomy and further strengthened astronomers' ambitions to observe further and better.

Not all of the early work done on Mount Wilson extended out to distant galaxies. Using specially mounted mirrors attached to the Hooker telescope, physicist Albert Michelson was able to take measurements of star sizes. The bright star Betelgeuse in the constellation of Orion was one of their targets, and those measurements contributed to theories about red supergiant stars and their origin and evolution. Stellar spectroscopy was a large part of the work done on the 100-inch telescope. It led directly to better determination of the chemical characteristics of stars and measurements of their distances.

Mount Wilson Today
Mount Wilson remains an active observatory, although its wonderful 'seeing' has been largely curtailed by the huge amounts of light pollution from the Los Angeles basin. However, that hasn't stopped the observatory from expanding its work. Currently, the mountain is also home to an optical interferometry array, the Center for High Angular Resolution Astronomy instrument, known as CHARA. Optical interferometry combines incoming light from two or more telescopes, in a manner similar to what radio astronomy arrays do for incoming radio signals from celestial objects. The advantage of this technique is that it lets astronomers 'simulate' a telescope of much larger aperture by using more than one telescope. Think of it

this way: an array of telescopes can provide an image of the same quality as a single telescope which would have the same diameter of all the telescopes in the array. The CHARA facility has six 1-metre (39-inch) reflectors ganged together. Each of the telescopes sends its own data to a facility where it is 'matched up' with imaging data from the others to provide a sharp, highly detailed image. This interferometry gives CHARA the same resolving power as a 330-metre (1,082-foot) telescope.

Beyond research, Mount Wilson Observatory offers stargazing nights for the public, as well as educational programmes for students and teachers throughout the Los Angeles area.

Creating a Golden Age

The spread of observatories from prehistory to the end of the 19th century set the stage for what many called the 'Golden Age of Astronomy' that began in the 20th century. Not only did observatories set forth on new paths of research, where they would expand our understanding of the cosmos, but they also served as sources of inspiration for a public hungry for news of the latest discoveries. Consider, for example, the Mars Mania that gripped the public when Percival Lowell talked about his search for Martians. He saw a chart of Mars made by the Italian astronomer Giovanni Schiaparelli, where a series of crisscrossing lines were labelled *canalis* – the Italian word for 'channel'. Lowell took the word to mean 'canals', which (to him) indicated structures built by intelligent beings. He wrote books about his Martian theories, and just as rumours about aliens and other oddities fly on social media today, his ideas about Martians gripped people's imaginations at the time. It was sensational news and the media ran with it.

Other astronomers felt Lowell was wasting his time, and were quite critical of his efforts. Yet, he spent many nights observing Mars, looking for evidence of Martians. Sadly, Lowell never found them and next turned to the search for the mysterious 'Planet X' of his. He applied mathematical calculations to determine there was 'something' out there. Lowell never found that either before he died, however, as previously mentioned, his obsession led to the hiring of Clyde Tombaugh, who did find the planet after all. His discovery caught the public imagination again, set against a backdrop of a new literary form called 'science fiction', which focused on tales of space travel and aliens.

Lowell's story wasn't the only astronomical tale to excite people. Events such as the Transits of Venus in 1874 and 1882 and recurring solar and lunar eclipses, also fed into people's growing fascination with the sky. Discoveries made by the up-and-coming crop of observatories and the continuing announcements about how observations of the solar eclipse of 1919 would prove Albert Einstein's prediction of gravitational lensing also fed into media sensationalism of science in general.

There's a valuable lesson in Lowell's story, and in the saga of astronomers traveling the world to see an eclipse: science can be exciting to people. If they have a chance to look through a telescope, many people come away impressed and perhaps even changed by what they see. That, in part, is why so many amateur astronomers exist. And it's also why in many places with university observatories, public viewing nights have been a staple for much of the 20th century.

4

THE TWENTIETH CENTURY AND BEYOND

The 20th century saw the rise of ever-larger and more complex telescopes and observatories around the world. On mountaintops in California, visionary astronomers built optical and infrared facilities that expanded not just humanity's view of the sky, but our understanding of the universe. As the century progressed, observatories moved equipment and labs to more and more remote areas to get better viewing conditions for optical, infrared, and radio astronomy. Places such as Kitt Peak in Arizona, the hill country of Texas, the summit of Mauna Kea in Hawaii became pre-eminent centres for astronomy by century's end. South of the equator, facilities sprang up at Mount Stromlo and Siding Spring in Australia, along with radio telescopes and arrays. World-wide consortia consisting of universities and government institutions banded together to put telescopes on Cerro Tololo, Cerro La Silla, and Cerro Pachón in Chile, and the Canary Islands in Spain. By the century's end, centres of astronomical and astrophysical exploration existed on nearly every continent, including Antarctica. And, telescopes were getting bigger. Larger ones in the 8-metre to 10-metre (26-foot to 32-foot) class comprised the 'next generation' of observatories poised to make major discoveries in the solar system, the galaxy, and beyond.

If someone wanted to tour all of the world's observatories, such a trip would include treks up mountaintops, visits to facilities in the high deserts, trips underground and underwater, flights high in the atmosphere, and out to space. It would be an amazing trip, and a lengthy one. The Minor Planet Center at the Smithsonian Astrophysical Observatory (part of the International Astronomical Union), lists more than 4,228 observatories in existence and reporting in 2019. This includes all the professional and amateur facilities, university sites, and observatories which are science centres or are connected to science centre/planetarium facilities. There are more observatory sites that exist, but not all are open; some are defunct and closed, others have been demolished or decommissioned (whether on the ground or on orbit).

Since it would take a volume twice as long as this book to talk about every observatory in the world, this chapter serves as a guided tour through a limited selection of ground-based observatories currently in use on the ground, their development, and what research questions they are being used to answer today. In Chapter 5, we will visit space-based observatories, and in Chapter 6, we'll take a look at the most unusual astronomy-related research facilities.

Northern Hemisphere Observatories

Our virtual observatory tour begins in California, where the first of the big, modern telescopes were built, starting late in the 19th century. In the last chapter, we first encountered George Ellery Hale, who was the founder of the California Institute of Technology (Caltech), building telescopes for Yerkes in Wisconsin and Mount Wilson near Los Angeles. He didn't stop there. He once boasted that he built the world's largest telescope not once, but four times. And, although he didn't realise it at the time, Hale was, along with Lowell in Arizona, stoking public interest in astronomy.

One spinoff of Hale's Mount Wilson Observatory construction, and a lesson in how valuable a public peek through a telescope can be, was the construction of the Griffith Observatory on Mount Hollywood in Los Angeles in the 1930s. This public facility was the brainchild of entrepreneur Griffith J. Griffith, who got interested in astronomy and made several visits to the newly built Mount Wilson in 1904. While there, he had a chance to look through the 60-inch telescope, which was then the largest telescope in the world. Apparently, Griffith was incredibly moved by what he saw and wanted to do something to bring astronomy to more people. Although he had already donated a great deal of property and money to the City of Los Angeles, in 1912, he offered the city the then-princely sum of $100,000 to build an observatory on Mount Hollywood. He wanted it to have the best astronomical telescope for public use, and he also suggested a public theatre where exhibits and motion pictures about astronomy could be shared. Griffith died before his observatory could be built. But, his dream lived on, and in 1930, planning for the facility began. His friend Hale contributed his vision to the building's design and the selection of a telescope. Physicists and technical experts from Caltech brought their expertise to the project. Exhibits were planned, including a Foucault pendulum and a model of the Moon. For sky viewing, the people in charge of the project selected a triple-beam coelostat for continuous solar viewing, and a 30-cm (12-inch) Zeiss refractor for night-time observations. Rather than put in a flat-screen movie theatre, the observatory's builders deviated from the original plan to include a planetarium in its place. They purchased a state-of-the-art Zeiss instrument to project stars onto a dome.

Upon its opening on 14 May 1935, Griffith Observatory became the property of the City of Los Angeles, with the stipulation the public would be able to visit for free to look through a telescope. Griffith's infatuation with astronomy echoes across the halls of his eponymous observatory, which was renovated and reopened in 2006. Its motto, 'Turning Visitors into Observers,' continues to send a message to visitors that the universe is ours to imagine and explore. Since its opening, nearly 10 million visitors have looked through its telescope and safely viewed the Sun, a testament to the interest in astronomy which exists in the public mind.

The Story of Palomar

George Ellery Hale's last great vision was made real on Palomar Mountain outside San Diego, California. The Palomar Observatory became the home of the largest telescope ever built up to that time, the 5-metre (200-inch is what it's commonly called) Hale Telescope. The observatory was completed in 1948 and rapidly became the most productive one of its time. Today, astronomers use it for everything from solar system studies to extrasolar planet searches, studies of stars, and the exploration of distant galaxies.

Palomar Observatory lies at the top of a roadway that winds through orange groves and farms in the region surrounding the mountain. Visitors change elevation rapidly, from near sea-level altitude to 1,871 metres (1.1 miles) at the summit. While it can be balmy and sunny at the foot of the mountain, winter weather can and does occur at the top, making road conditions treacherous. However, though much of the year, Palomar offers the same good viewing conditions which originally drew Hale and others to California in the first place.

The main dome and telescope were designed by noted American illustrator and Caltech staffer Russell Porter. Scientists from the Institute contributed engineering expertise for the mounting and dome construction and much of the infrastructure was created by companies from Los Angeles. It took nearly twenty years for Hale's dream to be accomplished, slowed by the Second World War and funding vagaries. Hale sought and received $6 million from the Rockefeller Foundation to build this observatory, but sadly, he did not live to see it completed and commissioned.

The components for the telescope itself were created by Westinghouse Company in Pennsylvania, and then shipped via the Panama Canal to San Diego. The telescope's building was finished first, while the mirror was created by the Corning Glass Works from a material the company had developed called Pyrex. It was a complex operation and two disks were created. The first one failed, but the second one was cast in December 1934. After the mirror blank cooled, it was shipped from Corning to the Caltech campus in Pasadena for grinding and shaping. It arrived April 1936, and took eleven-and-a-half-years for grinding. Part of the delay was due to the war and the retooling of precision equipment to turn out materials for military use.

Once it was ready for installation, the mirror made its way up to Palomar in November 1947, traveling along a road that was called 'The Highway to the Stars'. Once on the mountain, the mirror was tested, with additional polishing, grinding, and metal coating taking

The Hale 200-inch mirror during the grinding process, done at a special facility in Pasadena, California, before the mirror was transported to Palomar Mountain. Grinding was halted in 1942 due to the Second World War. The honeycomb support structure on the back of the mirror can be seen through the surface. Public domain image.

place over the next several months. The 200-inch Hale telescope was dedicated on 3 June 1948, designed for photographic work using glass plates in a specialised camera. Once the telescope was ready, astronomer Edwin P. Hubble took the first official images through the telescope. His target: a cloud of gas and dust called 'Hubble's Variable Nebula'. However, there remained some work to be done on the telescope, so official research didn't begin until early 1949.

The Hale Telescope retained its position as the largest telescope in the world until 1975, when the Russian Academy of Sciences built its BTA-6, which measured 6 metres (19.6 feet) across. The Hale was engineered, just as its sister telescopes in California were, to withstand earthquakes and deliver very steady views of the sky. The mirror and its associated instruments are held in place by an equatorial mount that allows for motion in all directions (east–west and north–south), so the telescope can cover large areas of the sky. Such a heavy telescope and mount requires several different motors so it can both move and track smoothly. Whenever instruments are installed on the telescope, the balance must be adjusted properly so the telescope's motions remain smooth. The telescope is anchored to mountain bedrock, using piers sunk 6.7 metres (21 feet) beneath the building. Not only does the telescope move, but the dome rotates as well. There are two circular rails for the dome, and two moving shutters to open the dome for observations.

Cameras and Instruments for Hale

Because the Hale telescope is a research-grade instrument, its use is scheduled for every clear night of the year. It is equipped with adaptive optics to correct for turbulence in the atmosphere, which gives its images a sharpness rivalled only by space-based telescopes. The telescope's scientific instruments include wide-field cameras, a large-format CCD camera, an imaging spectrograph, and optical and infrared spectrographs. The observatory continually upgrades its instrument capabilities, and is currently testing a high-speed camera and a stellar double coronagraph, which will allow it to look for extrasolar planets in the glare of their stars.

The 200-inch Hale isn't the only telescope on Palomar Mountain. There's also a 1.5-metre (59-inch) telescope, which expands the capabilities of the observatory. Joining it is the 1.2-metre (48-inch) Samuel Oschin Telescope (formerly known as the 48-inch Schmidt). This telescope was renamed after a prominent local entrepreneur who wanted to support the observatory. It has been used for numerous solar system studies, including the discovery of dwarf planet Eris. One of its most ambitious projects was the creation of an image called 'The Big Picture'. It is a single image that covers a tiny piece of the sky, about what you could cover with your finger held out at arm's length. Contained in this part of the sky is the central region of the Virgo Cluster of Galaxies, and a grouping called Markarian's Chain. The image was created using the QUEST camera attached to the Oschin Telescope. QUEST stands for Palomar Quasar Equatorial Survey Team Variability Study, which was a photographic study of the sky taken from 2003 to 2007. This camera uses a mosaic of 112 CCD detectors that cover a 4 x 4-degree field of view.

The first instrument to be used at Palomar was the 45-cm (18-inch) Schmidt camera (a catadioptric telescope) that came into service in 1936. It was used by astronomers such as Fritz Zwicky (who first suggested the idea of dark matter) to study supernovae in distant galaxies. Gene and Carolyn Shoemaker, with David Levy, discovered Comet Shoemaker-Levy 9, (formally called D/1993 F2) using this telescope in 1993. It's no longer used for research, but is part of a small display for visitors to the mountain.

From 2003 to 2009, the observatory also hosted a planet-searching instrument called the Palomar Planet Search Telescope. It was linked to similar telescopes in the Canary Islands and at Lowell Observatory until its decommissioning. Also hosted on the mountain was the Palomar Testbed Interferometer. While it was operational from 1995 to 2008, it allowed astronomers to perform high-resolution astrometry (the positions of objects in space), and measure the shapes of some of the brighter stars.

A Survey at Palomar

Most astronomers are familiar with the Palomar Observatory Sky Survey (POSS) that took place under the sponsorship of the National Geographic Institute and encouraged by Caltech astronomer Fritz Zwicky. The survey was created using what became the Samuel Oschin telescope to take photographic plates of the sky from the celestial north pole (90 degrees) to -27 degrees in the southern hemisphere. Thanks to the high sensitivity of the films used and the resolution provided by the Schmidt/Oschin telescope, the plates showed objects as faint as 22nd magnitude. The first survey was completed in 1958, and a second one was done in late 1958, which allowed it to be extended to the far southern latitudes. For many years, prints and negatives made from these plates were the gold standard for sky survey work, and were used in observatories around the world. (The author used a set during research into determining accurate comet positions in the 1990s.) The POSS has since been expanded, and in 1983, the Space Telescope Science Institute used the plates to create what became the Digitized Sky Survey (DSS). A second survey was performed beginning in the 1980s.

Today, Palomar Observatory continues as a centre for astronomy studies, despite issues it faces from light pollution from nearby cities. Using state-of-the-art techniques, astronomers are using it for observations across the universe, particularly for objects in the most distant reaches of the solar system, as well as gamma-ray bursts, quasar studies, and other topics.

Astronomy from the US Southwest

As we saw earlier, the southwestern state of Arizona in the United States provided a home for Lowell Observatory at the turn of the 19th century. When Lowell first set foot off the train that brought him east from Boston to Flagstaff, he was in the vanguard of astronomers seeking clear, dry conditions for optimal observing.

By the middle of the 20th century, in response to interest from astronomers, the National Science Foundation chose Kitt Peak, outside Tucson, as the site for a national observatory. It was placed on property leased from the Tohono O'odham nation, and today hosts twenty-two optical telescopes and two radio facilities. The Kitt Peak National Observatory (KPNO) is administered by the National Optical Astronomy Observatory (NOAO), which also oversees facilities in New Mexico, as well as the Cerro Tololo Inter-American Observatory in Chile. As with other major observatories, this site hosts researchers and students from around the country as well as scientists from selected other countries. Kitt Peak is a sought-after tourist site in Arizona, as well as a productive and busy observatory reserve. It is more than an hour's drive from Tucson, on a road winding through scrub desert country before winding up the mountain, to an altitude of 2,096 metres (6,876 feet).

The largest optical telescopes at KPNO are the Nicholas U. Mayall 4-metre (13-foot) telescope and the WIYN 3.5-metre telescope. Construction of the Mayall telescope began in 1968 and it saw its first light on 27 February 1973. The telescope's primary mirror weighs 15 tons and is covered with a thin, reflective aluminium coating only a thousandth the thickness of a human hair. The Mayall telescope is equipped with a CCD Mosaic camera that allows it to take very high-resolution images of distant objects. The telescope is used mainly for optical and infrared viewing, for such studies as rotation curves of distant galaxies and the structures of elliptical (egg-shaped) galaxies.

(Rotation curves are a measure of the variation in velocity of stars and gas clouds in a galaxy depending on their distances from the centre.)

The WIYN Observatory (operated by a consortium made up of the University of Wisconsin-Madison, Indiana University, Yale University, the University of Missouri, Perdue University, and NOAO) has been operating on the mountain since 1994. Its telescope has three mirrors, and the primary mirror is shaped by an active optics system. The observatory maintains a suite of instruments to attach to the telescope that enable studies in a wide range of areas, including asteroid observations, exoplanet searches, and other solar system bodies, as well as distant nebulae and galaxies.

One of the most striking-looking facilities on Kitt Peak is the McMath-Pierce solar telescope. Its slanting heliostat shaft appears to lean up against a tower. The main telescope is a 1.6-metre (5.2-foot) reflecting solar telescope, and has been the largest such facility in the world. Currently, it is closed and plans are underway to transform it into a visitors' facility.

Steward Observatory

The University of Arizona in Tucson is home to the Steward Observatory, where astronomy research has been done since 1916. The first major telescope was built on campus using a donation of $60,000 from Mrs Lavinia Steward. It paid for the construction of a 91-cm (36-inch) telescope, and after delays caused by The First World War, the telescope was installed in July 1922 and the observatory dedicated to the memory of Mrs. Steward's late husband in April 1923.

Steward Observatory operates facilities at three different locations in the Arizona Mountains: Mount Graham International Observatory, Mount Lemmon Observatory and Catalina Station on Mount Bigelow. It also has installations at Kitt Peak and the Fred Lawrence Whipple Observatory on Mount Hopkins, plus radio telescopes at Kitt Peak and Mount Graham. Steward Observatory operates the Richard F. Caris mirror laboratory on campus, where its technical personnel have cast and ground mirrors for such observatories as the Giant Magellan Telescope and the Large Synoptic Survey Telescope being installed in Chile. The Hubble Space Telescope and Spitzer Space Telescope, as well as the upcoming James Webb Space Telescope, have benefitted from work done at Steward. The HST Near Infrared Camera and Multi-Object Spectrograph, the Multiband Imaging Photometer on-board Spitzer, and the Multiband Imaging Photometer onboard Spitzer were also designed or built at Steward.

Large Binocular Telescope

Several world-renowned observatories sit atop Mount Graham. The Large Binocular Telescope (LBT) is one of the more unusual looking telescopes in the world. It sits at an altitude of 3,200 metres (10,498 feet). The telescope is operated by and for a range of partners around the world, including research institutions and universities in Arizona, Indiana, Minnesota, Ohio and Virginia in the US as well as in Germany and Italy. The LBT is really a pair of 8.4-metre (27-foot) mirrors trained on the sky just like a pair of binoculars. With that configuration, astronomers get the same field of view as a single mirror measuring 11.8 metres (38.7 metres) across. It has been in use since its first light with both mirrors on 11 and 12 January 2008. The observatory uses an adaptive optics system that deforms the secondary mirror inside the telescope. With this system, some of LBT's observations have been sharper than what Hubble Space Telescope can achieve. The observatory is equipped with cameras sensitive to both optical and ultraviolet light, spectrographs, and other imagers for interferometry.

VATT and SMT

The Vatican City State has partnered with the Steward Observatory to operate the Vatican Advanced Technology Telescope (VATT) on Mount Graham. This facility is used mainly for imaging and photometry (measuring the intensity of light from distant objects). The heart of the VATT is a 1.8-metre (5.9-foot) instrument called the Alice P. Lennon Telescope that primarily observes in the optical and near-infrared ranges of the electromagnetic spectrum. The mirror was created and polished at the Steward Observatory workshops and the secondary was cast at the Space Optics Research laboratory in Chelmsford, Massachusetts. Astronomers have used this facility for a variety of investigations, including the search for MACHOs (massive compact halo objects) orbiting in the Andromeda Galaxy.

Finally, the Submillimetre Telescope (SMT) on Mount Graham is a radio facility housing a 10-metre (32.8-foot) dish. It is operational throughout most of the year and maintained by the Arizona Radio Observatory.

New Mexico Astronomy

New Mexico, like its neighbouring states Arizona and Texas, is a rugged region with deserts and mountain ranges. Dry weather conditions there encourage astronomy observations, and the state boasts a number of well-known professional observatories and dark sky preserves.

The most famous of the New Mexico observatories is the Karl G. Jansky Very Large Radio Array (VLA), near Socorro, that stretches across a large flat region called the Plains of San Agustin. It's run by the National Radio Astronomy Observatory (NRAO), which operates the Atacama Large Millimetre Array in Chile and the Very Large Baseline Array in North America. Visitors can see the 25-metre (82-foot) dishes of the VLA across great distances before arriving at the heart of the array, and it's even visible from space. The VLA, unlike optical ground-based telescopes, is not sensitive to light pollution, but it can be affected by electronic pollution. For that reason, visitors to the array are asked to turn off their mobile phones and any other devices which might interfere with the array's sensitive radio dishes. The array was made famous in the movie *Contact,* starring Jodie Foster as a radio astronomer who detects what she thinks is an alien signal while using the VLA for observations.

The array has been in operation since the 1970s when it was built and commissioned. It has been upgraded over the years, and expanded to become part of the Very Long Baseline Array (VLBA) stretching from St Croix in the US Virgin Islands to Mauna Kea on the Big Island of Hawaii. The Event Horizon telescope, which was used to study the region around a black hole in the galaxy M87, used part of the VLBA array for its detection.

The VLA is used in different configurations to accommodate the various needs of each observation run. This means that several times a year, the telescopes are arranged in different configurations. Technical staff do this by loading each dish onto a railroad car and moving it to the proper place. Throughout the year, visitors can see the dishes in different places, depending on the types of objects being observed.

Science with the VLA

The VLA typically focuses in on very small areas of the sky to get detailed information about a wide range of events and objects. It has studied stars turning on, and others dying in massive supernova and hypernova explosions. It has regularly peered inside giant clouds of gas and dust to track radio emissions from the actions of newly forming stars hidden within. It provided very clear radio imaging of the black hole at the centre of our own Milky Way, an object astronomers dubbed 'Sagittarius A*' (pronounced 'Sagittarius A-Star'). It has even been used to study radio frequencies emitted by some very common precursor pre-biotic molecules found here on Earth that also exist in interstellar space.

Solar Viewing

As one might imagine, New Mexico is well-suited for solar observing. The best-known solar observatory is at Sunspot, New Mexico, at an altitude of 2,818 metres (9,245 feet or 1.7 miles). The first solar telescope was called the Grain Bin Dome and built in 1951. Later on, the John W. Evans Solar facility began operation, using a coronagraph. These two facilities, plus another one called the Hilltop Dome, were in use for many years. Today, only the Richard B. Dunn Solar telescope is in use. It was built by the United States Air Force in 1969 and originally called the Vacuum Tower Telescope. It was transferred to the National Solar Observatory, which operated it until 2017. Today, the observatory is under the administration of the Sunspot Solar Observatory Consortium, including New Mexico State University, the Associated Universities for Research in Astronomy, and the National Science Foundation.

Cataloguing the Universe

Apache Point Observatory, also located in Sunspot, is operated by New Mexico State University and operated by the Astrophysical Research Consortium (made up of the Institute for Advanced Study at Princeton, the University of Chicago, University of Washington, University of Colorado, University of Virginia, and Georgia State). This private facility maintains the ARC 3.5-metre (11.5-foot) telescope, plus a full suite of cameras, spectrometers, and spectrographs, the NMSU

1.0-metre (3.2-foot) telescope, and a small-aperture photometric telescope. In addition, Apache Point is the home of the Sloan Digital Sky Survey (SDSS). It uses a 2.5-metre (8.2-foot) optical telescope to capture images and spectra across a portion of the sky to study the redshifts of objects in its field of view.

The Sloan Digital Sky Survey, named for the Alfred P. Sloan Foundation, began in 2000 and continues to gather spectral information about millions of objects in its field of view. The resulting data is available for further study, and previous releases have been used in planetarium software, Google Sky, and Microsoft's World Wide Telescope (now administered by the American Astronomical Society). The SDSS is currently undertaking its fourth major survey of galaxies in both the northern and southern hemisphere skies.

Astronomy in Texas

This state hosts two large astronomy research telescopes at the McDonald Observatory atop Mount Locke in the mountains of west Texas. The facilities are administered by the University of Texas in Austin. The observatory was founded in 1939, named after Texas banker William Johnson McDonald, and at the time of its dedication it had the world's second-largest telescope. The heart of the observatory is the Otto Struve Telescope, a 2.1-metre (82-inch) instrument. It was later joined by the 2.7-metre (8.8-foot) Harlan J. Smith Telescope, and the 10-metre (32-foot) Hobby-Eberly Telescope (HET) on nearby Mount Fowlkes. The HET is optimised specifically for spectroscopic studies of the sky, and is operated by a consortium consisting of the University of Texas, Penn State University, and the Ludwig Maximilians University in Munich and Georg-August University of Göttingen. Such multi-institution cooperation is a hallmark of modern astronomy, allowing maximum science to be done through cost- and talent sharing.

The HET, which is one of the largest of the modern optical telescopes, doesn't have just one primary mirror. Instead, it has one large reflecting surface that is made of 91 hexagonal segments. To achieve high-resolution studies of galaxies, solar system objects and the search for exoplanets, the HET is equipped with three spectrographs, and the telescope's mirrors can be adjusted by actuators under each segment.

Canadian Observatories

Astronomy in Canada was largely a frontier undertaking until about 1750, when the first purpose-built astronomy observatories were built. Before then, explorers noted comets and eclipses, and used star positioning to guide them as they trekked across the country. Among the earliest-known observation posts was a site built at Louisbourg by the Marquis de Chabert. There, he observed the skies and created star maps. In 1765, Joseph Frederick Wallet des Barres, a surveyor, built an observatory at Castle Frederick, Nova Scotia. He apparently used it to test his surveyor's instruments and may well have had telescopes there, as well. Nothing remains of these two early facilities.

Interest in astronomy and stargazing in Canada proceeded along similar lines of research as in other countries. At least one observatory was built on the Citadel in Quebec City in 1840, and a successor facility on the Plains of Abraham in 1874. Both were aimed at navigational needs rather than scientific observations. In particular, as we discussed elsewhere, determining longitude for ships at sea or in newly explored lands was a challenge for sailors and explorers. One method was to use eclipses, which early Jesuit explorers did in Canada, and others did in Europe. British clockmaker John Harrison finally invented a practical marine chronometer in 1761, which solved the problem. Interest in the sky continued, and observatories were created around the country.

The first purpose-built astronomical observatory appeared at the University of New Brunswick in 1855, followed by the unveiling of a public observatory in Kingston. In 1862 an observatory that supported timekeeping activities was built in Montreal. As the country grew, other towns and cities built observatories. When the country's transcontinental railway was built, the government looked to observatories in remote regions to assist with land surveying and navigation.

In the 18th and 19th centuries, two types of events occurred which sparked further interest in astronomy in Canada. Eclipses, which had been of interest to those trying to calculate longitudes, became more of a public spectacle and events of scientific interest. As almost every eclipse-chaser knows, eclipses don't always take place in regions that are easily accessible. Today, it's easy enough to fly somewhere to get in the path of an eclipse, but in earlier centuries, a trip to see a total solar eclipse was a major expedition. Still, they provided interesting science about the Sun, and so people trekked across Canada to see such spectacles as the 17 July 1860 totality. Another, much rarer event than an eclipse, is a transit of Venus. These are seen from Earth every 243 years, and happen when the planet Venus crosses in front of the Sun during its orbit, as seen from Earth. As interest grew in charting solar system motions, transits of Venus became one way of calculating distances between planets. The transit of 1769, which excited astronomers around the world, drew visitors to Canada to observe it. Since transits repeat in pairs, another one was due to occur in 1882, and again, astronomers came to Canada to view it.

Astronomical research really took off in Canada beginning in the 20th century. The government built the Dominion Observatory in Ottawa in 1902 to serve both astronomical and timekeeping needs. It was outfitted with a refracting telescope and a reflecting solar telescope, and also performed positional astronomy measurements using transit instruments. A few years later, in 1918, the Dominion Astrophysical Observatory opened near Victoria, British Columbia. It boasted a 1.88-metre (73-inch) telescope that was, for a short time, the world's biggest telescope. This observatory is still in operation and has been modernised several times. Facilities at Victoria have been upgraded with the addition of a second telescope and continuous modernisation of detecting equipment.

The David Dunlap Observatory was built outside of Toronto 1935 by the University of Toronto for its astronomy students. It contains a 1.88-metre (74-inch) reflector, and several other instruments including a small radio telescope. In 2008 the observatory was taken over by the Royal Astronomical Society of Canada, an organisation of amateur astronomers committed to advancing astronomy.

The second half of the 20th century saw a rise in observatory building across Canada. This included the creation of a set of radio telescope observatories in Penticton, British Colombia, and at Algonquin Provincial Park (which is now closed). The Penticton facility, called the Dominion Radio Astrophysical Observatory (DRAO), was built in 1960 and expanded to an array of telescopes. Today, there are seven 8.5-metre (27.8-foot) parabolic antennas at the site and they focus on solar activity and were used on the Canadian Galactic Plane survey from 1995 to 2005. The DRAO is now under the management of the National Research Council and its Herzberg Institute of Astrophysics.

In 1995, the DRAO and a consortium of astronomers in North America and Europe, began a project to use radio telescopes to survey emissions from atomic hydrogen and other sources in the interstellar medium of the Milky Way Galaxy. The resulting work is called the Canadian Galactic Plane Survey, and to date has provided very high-resolution mapping of the gas and dust between stars in our galaxy.

Canada is also part of the various consortia that built the Canada-France-Hawaii telescope, the Gemini Observatory, the James Clerk Maxwell Telescope (all located on Mauna Kea on the Big Island, Hawaii, and Gemini South in Chile). It is also a partner in the Atacama Large Millimetre Array in the Andes.

The Mirrors of Mauna Kea

The rise of infrared astronomy, and the need for places on Earth where astronomers can get good 'seeing' and detect the infrared light that *can* make it through our atmosphere, played a role in the establishment of astronomy reserves in the late 20th Century. The Mauna Kea Observatories sit high atop a dormant volcano on the Big Island of Hawaii at an altitude of 4,205 metres (13,795 feet) above sea level and usually well above the clouds that cover the rest of the island. The drive to the observatories affords visitors incredible views of the nearby volcanoes of the Big Island, and the surrounding Pacific Ocean. It's not a drive for the faint-of-heart, since the road is made largely of pulverised volcanic rock which has to be graded frequently and drivers must watch their speed on the hairpin turns. Once they arrive, visitors and working astronomers are surrounded by what seems like another realm: a caldera filled with cinder cones, very little vegetation, and crowned by a collection of observatories. There, astronomers have built and installed a collection of thirteen facilities. Nine are optical or infrared-enabled facilities, while three work in the submillimetre range, and one in the radio spectrum.

What makes Hawaii, and Mauna Kea in particular, such a useful astronomy site? For one thing, it's isolated in the Pacific Ocean, and the humidity at altitude is very low, the air is thin, and the trade winds kept things dry. This makes the top of the mountain a perfect place for infrared viewing, as well as optical astronomy.

The first telescope on Mauna Kea was installed by astronomer Gerard Kuiper in the 1960s. By 1967, the University of Hawaii astronomy department had won a grant to build a 2.24-metre (88-inch) telescope at the summit. First light for the telescope, often called the 'UH 88', occurred in 1970.

Due to its opening, word got out that the mountain was a good place to do astronomy. Over the next few decades a number of observatories were built, including the Canada-France-Hawaii facility, equipped with a 3.6-metre (141-inch) telescope, the Gemini North facility, the Subaru Telescope (managed by the National Astronomical Observatory of Japan), the twin Keck telescopes, each of which features 36 segmented mirrors, the Caltech Submillimetre Observatory, the NASA Infrared Telescope Facility, the James Clerk Maxwell Telescope, a Submillimetre array, the United Kingdom Infrared Telescope, and one part of the Very Long Baseline Radio array. The combined telescopes on the mountain represent investments and management from ten countries and an array of universities and research institutions. The astronomy reserve is overseen by the Office of Maunakea Management, which is responsible for daily management and enforcement of the mountain master plan.

Gemini: Two Telescopes, One Observatory

Gemini Observatory is one of the newer facilities on the mountain and a good example of the type of infrared- and optical-light sensitive research institution on Mauna Kea. It is actually an observatory with two components: one on Mauna Kea and the other on Cerro Pachón, in Chile. Each of the twins has optics optimised for both optical and infrared viewing, and each has a suite of instruments including cameras, spectrographs, and adaptive optics installations. Between the two telescopes, Gemini provides nearly 100 per cent coverage of both the northern and southern hemisphere skies. Each one is operated from a base facility – in Hilo, Hawaii, for Gemini North, and La Serena, Chile, for Gemini South. Astronomers can come to the summit for their observations, although most astronomers simply have the observatory execute their observing programmes and then gather the data when it's ready for analysis.

Both Gemini telescopes were planned and built on their remote mountains to provide clear seeing without the atmospheric distortion that plagues telescopes at lower altitudes. Together, they cost an estimated $184 million to build and commission. A night at Gemini costs tens of thousands of dollars, covered by funding agencies which support the observatory.

Each telescope's main mirror is 8.1 metres (26.5 feet) across and is very thin compared to older telescope mirrors. Each 20-cm (7-inch)-thick mirror is covered with a thin silver layer that enables its visible-light and infrared viewing. Both were fabricated at the Corning glass works in New York. These are flexible reflectors which each lie on a 'bed' of 120 actuators that gently shape them for astronomical observations.

The second of the Gemini twin telescopes is located on Cerro Pachón, in the Chilean Andes Mountains. It's up at 2,700 metres (8,850 feet). Like its sibling in Hawaii, Gemini South takes advantage of very dry air and good atmospheric conditions to observe the southern hemisphere skies. It was built about the same time as Gemini North and made its first observations (called first light) in 2000.

The twin Gemini telescopes are outfitted with a number of instruments, including a set of optical imagers, plus spectrographs that dissect incoming light into its component wavelengths. These instruments provide data about distant celestial objects that are not visible to the human eye, particularly near-infrared light. The observatory's targets are planets, asteroids, clouds of gas and dust, galaxies, and galaxy clusters.

One particular instrument, the Gemini Planet Imager, was built to help astronomers search out extrasolar planets around nearby stars. It was first used at Gemini South in 2014. The imager itself contains a coronagraph, spectrograph, adaptive optics, and other parts that help astronomers locate planets around other stars. It has been continually tested and improved and one of its most successful planet searches turned up the world 51 Eridani b, which lies about 96 light-years away from Earth.

Since Gemini opened, it has peered into distant galaxies and studied the worlds of our own solar system. Among its most recent discoveries, astronomers used Gemini North to look at a distant quasar (an energetic galaxy) which had previously been observed by two other observatories: the Keck 1 on Mauna Kea and the Multiple-Mirror Telescope (MMT) in Arizona. Gemini's role was to focus on a gravitational lens that was bending the light from the distant quasar toward Earth. Gemini South has also studied distant worlds and their actions, including one which may have been kicked out of orbit around its star.

Other images from Gemini include a look at a colliding galaxy called a polar ring galaxy. This one is called NGC 660, and the image was taken from the Fredrick C. Gillett Gemini North telescope in 2012. More recently, Gemini has been used to look at metal-poor stars, and to make detailed studies of a jet streaming away from a central supermassive black hole in a distant elliptical galaxy.

Double Eyes on the Sky: The Keck Telescopes

Visitors taking a tour of the Mauna Kea observatories can often get a chance to see the twin Keck telescopes. The W.M. Keck Observatory and its two 10-metre (32-foot) wide telescopes are sensitive to optical and infrared light, and are among the world's largest and most productive instruments. The Keck mirrors are each made of 36 hexagonal segmented mirrors made of a 7.6-cm (3-inch)-thick glass-and-ceramic composite called Zerodur. This material resists heat and cold, but is not reflective on its own. Each of the mirror segments is coated with a very thin layer of aluminium. Even though it reflects light well, the mirror coating must be cleaned often and replaced when it loses its reflectivity. As with other modern observatory facilities, the Kecks are outfitted with an array of spectrographs, along with infrared cameras. In recent years, the observatory has also installed adaptive optics systems that help its mirrors compensate for the movement of the atmosphere which can blur the view. The adaptive optics lasers help measure the atmospheric motions, and then the rest of the system helps correct for turbulence by sending instructions to a set of actuators on a deformable mirror that changes shape 2,000 times per second. The Keck II telescope became the first large telescope worldwide to develop and install an adaptive optics system in 1988 and was the first to deploy lasers in 2004.

The history of the Keck observatory stretches back to the early 1970s, when astronomers began planning a new generation of large ground-based telescopes with the largest mirrors they could create. However, glass mirrors can be quite heavy and ponderous to move. What the scientists and engineers wanted were light-weight ones. Astronomers at the University of California and Lawrence Berkeley Labs were working on new approaches to building flexible mirrors, and they came up with a way to do it by creating segmented mirrors that could be angled and 'tuned' to create one larger mirror. The first mirror went into Keck and had its first light in May 1993. Keck II opened in October 1996. The cost to build and commission these facilities was about $192 million, plus additional costs of about $78 million for the instruments. Today, one night at Keck costs nearly $60,000 dollars, and which is funded through the National Optical Astronomy Observatories consortium, and the US National Science Foundation.

Currently, the observatory is used not only for astronomical observations, but also to support spaceflight missions. It will also be used in conjunction with the upcoming James Webb Space Telescope. The W.M. Keck Observatory is managed by a consortium called the California Association for Research in Astronomy (CARA), which includes cooperation with Caltech and the University of California. NASA is also part of the partnership. The W.M. Keck Foundation provided funding for its construction.

The Keck telescopes are in great demand for a variety of optical and infrared studies. More than 25 per cent of observations made by US astronomers are done at Keck, and with its improved instruments, and adaptive optics, the observatory can often turn out images that approach and sometimes surpass the view from the Hubble Space Telescope. Keck's wide range of observations in visible and near-infrared light opens up a realm of interesting objects to astronomers. Among them are star birth regions, where young stellar objects are just starting to form. It detects infrared light coming from inside the stellar nursery to see what's happening there. Keck is also used to study hot young stars already born, which energise the clouds of material that formed their 'nests'.

Keck isn't limited to star birth and hot young objects. Its twin telescopes have been used to observe objects at the infancy of the universe. In one case, they looked at an extremely distant cloud of gas which existed shortly after the birth of the universe, some 13.8 billion years ago. This distant clump of gas isn't visible to the naked eye. However, using specialised instruments, astronomers observed a very distant quasar as its light passed through the cloud. From the Keck data, astronomers discovered that the cloud was made of pristine hydrogen. This means it existed at a time when other stars had not yet 'polluted' space with their heavier elements. It's a look at conditions back when the universe was only 1.5 billion years old.

Another question Keck-using astronomers want to answer is 'how did the first galaxies form?' The most current theory is that the 'shards' of the first ones were likely coming together around 400 million years after the Big Bang. This puts them at a very great distance. Since they are very far away from us, observing those early galaxies can be difficult since they appear very dim. In addition, their light has been affected by the expansion of the universe and what was originally emitted in ultraviolet, for example, now appears in the infrared part of the spectrum. Understanding infant galaxies can supply a probe into early cosmic history, and they help astronomers see how our own Milky Way

A 2005 view of the Subaru Telescope, left, the W.M. Keck twin telescopes, center, and the NASA Infrared Telescope Facility, right, on Mauna Kea in Hawaii. Courtesy of Sasquatch, Creative Commons Attribution-Share Alike 3.0 licence.

formed. Keck can observe those distant early galaxies with its infrared-sensitive instruments. Astronomers can then study the light being emitted by hot young stars in those galaxies (their light was originally emitted in the ultraviolet), which is re-emitted by clouds of gas surrounding them. The infrared signals that arrive at Keck's mirrors give astronomers some insight into conditions in those distant stellar cities at a time when they were mere galactic infants, just starting to grow.

The Subaru Telescope

Not far from the Keck and Gemini Telescopes is Japan's contribution to astronomy on Mauna Kea: the $377 million Subaru Telescope. It's named after the Japanese name for the Pleiades star cluster. Subaru was commissioned in 2005. At that time, its 8.2-metre (26-foot) mirror was the largest single telescope mirror in the world.

Planning for Subaru began in 1984 at the University of Tokyo. After several years of planning, the University signed a contract with the University of Hawaii allowing the construction of the telescope on Mauna Kea. Today, Subaru Telescope is managed through the National Astronomical Observatory of Japan. Like Gemini, the Subaru mirror is shaped by a collection of actuators which gently guide it to the proper configuration. Also like other modern observatories, Subaru is complemented by a suite of instruments and an adaptive optics system, including the state-of-the-art Hyper Suprime-Cam. This 900-megapixel ultra-wide field of view camera has been in use since 2012.

Subaru Science

Due to its high sensitivity and the high resolution of its instruments and cameras, the Subaru Telescope is much in demand for studies of dim, distant objects in the early universe. For example, in early 2019, astronomers using Subaru detected 83 quasars powered by central supermassive black holes. These intensely bright objects date back to the very earliest epochs of the universe's 13.8-billion-year history. Their existence raises questions about how black holes formed and grew in the early universe.

Subaru has also been making detailed observations of a young planetary system about 500 light-years from Earth. Preliminary analysis shows there could be up to three planets around a young, Sun-like star called LkCa 15. More recently, Subaru has been used to study about 1,800 supernovae in the distant universe. These exploding stars shine incredibly brightly for a short time, and astronomers often use them as a so-called 'standard candle' to help calculate distance measurements across the universe. The death stars Subaru found are called superluminous supernovae, and they appear as they existed in the very early universe. Studying their light can give more information about those early cosmic times, and the process of star birth among the most massive stars that existed then.

Submillimetre Astronomy from Mauna Kea

Optical and infrared astronomy are not the only electromagnetic spectrum 'regimes' that are observed from Mauna Kea. There are also facilities for submillimetre science, including at least one array, and the odd-looking James Clerk Maxwell telescope (JCMT). It's situated in a valley beneath Keck, Subaru, and Gemini, and was built and managed by a joint consortium of facilities in the United Kingdom, the Netherlands, and Canada. In 2015, the facility was taken over by another consortium, the East Asian Observatory. The JCMT has, in the past has been linked with the Caltech Submillimetre Array next-door to form an interferometer.

The heart of the JCMT is a 15-metre (nearly 50-foot) primary reflector made of 276 individual panels, each skinned with a thin layer of aluminium. Every panel can be adjusted to help maintain a parabolic shape for the telescope's entire reflective surface. The whole assembly is on a carousel that rotates, and hidden from the elements by a membrane which protects the telescope from sunlight, rain, and wind.

What is Submillimetre Astronomy?

The submillimetre range of light lies between the far-infrared and microwave portions of the spectrum. Until relatively recently, this remained a poorly studied region of light, despite it being so close to the radio and infrared range. That's partly because submillimetre observations can be difficult, particularly due to of Earth's atmospheric density, which interferes with incoming submillimetre frequencies.

So, just as infrared observing is done best at high altitudes (or from space), submillimetre astronomer benefits from being done high up on Mauna Kea and at other high-altitude sites such as the Chajnantor Plain in Chile.

Astronomers using submillimetre arrays and telescopes generally tend to zero in on molecular clouds (usually made up of molecules of hydrogen and other gases), where star formation is likely to happen. They can also perform astrochemistry by looking at the gases and dust that make up those clouds.

Distant galaxies also undergo episodes of star formation, and this activity gives off wavelengths of light ranging from ultraviolet to microwave. By ganging together submillimetre telescopes to make interferometric observations, astronomers can study the distant star-forming clouds in other galaxies to compare and contrast with the same activities we see in the Milky Way. Finally, as with Subaru and other observatories that focus on infrared light from very distant objects in the young universe, submillimetre telescopes can offer another look at their origins and evolution. In particular, such observatories can continue studies of the cosmic microwave background radiation, which dates back to a time when the hot young universe finally became transparent enough to allow light to travel freely.

The Future of Mauna Kea
The Mauna Kea Observatories Reserve is one of the world's premier observing spots, and will remain that way for some time to come. However, it does operate under some limitations. Currently, there are only thirteen working facilities allowed at the summit of the mountain. This is due to agreements with the native Hawaiian population, which considers the top of the mountain to be sacred ground. The future Thirty Meter Telescope (TMT), which is slated to be built on Mauna Kea, came under intense scrutiny and its construction has been met with opposition among a minority of native Hawaiians. At the time of writing, the TMT was approved to move ahead. Part of the agreement states that at least two facilities currently on the mountain must be decommissioned. However, protests against its construction continue. We will look at this facility in more detail in Chapter 7.

Astronomy in China

In Chapter 2, we had a brief look at the history of astronomy in China. Although the country has had a long history of stargazing, dating back to prehistoric times, it wasn't until the early 20th century that Chinese astronomers began building significant observatories for scientific examination of the sky. One of the best known of the early 20th century Chinese astronomy institutions is the Purple Mountain Observatory in Nanjing. It was formerly part of the Institute of Astronomy, which constructed the observatory in 1934, and it became known as the 'Cradle of Modern Astronomy in China'.

For a time, Purple Mountain had the largest telescope in the Far East, a 60-cm (23.6-inch) reflector. After the Second World War, observations resumed and the observatory expanded. Over the years, the Purple Mountain Observatory and its astronomers have been credited with the discovery of 149 minor planets. It was also active in near-Earth object asteroid studies from 2006 and 2013, and found more than 600 objects. Due to encroaching light pollution, however, much of the current research done under the auspices of the Purple Mountain Observatory is now being done at five additional observatories in Ganyu, in Eastern China; Honghe, abutting Russia; Qingdao, on the Yellow Sea coast;, and Qinghai, on the Tibetan plateau.

Other observatories in China include the Beijing Astronomical Observatory, which maintains two facilities in Xinglong and Miyun. The Xinglong observatory is outfitted with a 2.16-metre (85-inch) reflector telescope, which is China's largest, and a 1.26-metre (49-inch) infrared-sensitive telescope. Miyun is the home of the Beijing Observatory's 28-dish radio array – the Metre-Wave Aperture Synthesis Radio Telescope. In recent years, the Beijing Astronomical Observatory has been merged with several other Chinese facilities to form the National Astronomical Observatory of China.

The most recent large observatory project in China resulted in the creation of the Five-hundred-meter Aperture Spherical Radio Telescope (FAST), also known in Mandarin as the 'Eye of Heaven'. This huge radio telescope, made of 4,450 triangular panels fitted together to make a single reflecting dish, is located in a natural depression in the Guizhou Province of south-western China. For the moment, it is the world's largest filled-aperture radio telescope and the second-largest single-dish radio telescope. (The largest is in Russia, the RATAN-600.) During its testing and commissioning phases, FAST discovered two new millisecond pulsars. These are old neutron stars that are spinning on their axes in a few thousandths of a second. The observatory also observed gamma-ray pulsars, and is expected to make major contributions to understanding gravitational waves, cosmic rays, and to probe the nature of the interstellar medium. Future plans also include a deep-sky survey at radio wavelengths. The observatory is sensitive to frequencies ranging from 70 MHz to 3.0 GHz.

While FAST is expected to be a very productive observatory, scientists are having to deal with an influx of tourists, intent on seeing this huge telescope set into a sinkhole. With tourists come cell phones and other forms of radio frequency interference.

Japanese Astronomy

Japan has a rich cultural history dating back to prehistoric times. Like many other cultures, the Japanese interest in the sky was partly practical – to use astronomical observations to calculate the calendar, as a timekeeping practice, and to determine correct longitude and latitudes for navigation. The country's participation in scientific aspects of astronomy didn't really begin until the late 1700s. That was when the Asakusa Observatory was created under the Shogun in the late Edo era. It wasn't until 1888 that the Tokyo Astronomical Observatory (TAO) was established, and Hongo Campus of the University of Tokyo built an observatory for training students. The TAO was moved to Mitaka in 1924 and a 65-cm (25.5-inch) refracting telescope was installed. Later on, a solar tower telescope was built.

The era of modern astronomy in Japan began in the 20th century, when the Tokyo Astronomical Observatory and other facilities were all reorganised under the NAOJ, the National Astronomical Observatory of Japan. On 1 April 2004, the NAOJ fell under the auspices of a reorganised institution that became known as the Inter-University Research Institute Corporation, National Institutes of Natural Sciences, National Astronomical Observatory of Japan.

The 1950s began an era of telescope building that resulted in the Norikura Solar Observatory, the Okayama Astrophysical Observatory, the Dodaira Observatory, and the Nobeyama Solar Radio Observatory. The largest project built by NAOJ was the Subaru Telescope, on Mauna Kea on the Big Island of Hawaii (as mentioned earlier).

In recent years, NAOJ has worked in partnership with a number of other organisations around the world. For example, it is part of the ALMA consortium, and is placing telescopes in Chile, as well as in Australia and across Japan. The NAOJ is a major partner in the Thirty Meter Telescope, to be built on Mauna Kea. It is also building the Large-Scale Cryogenic Gravitational Wave Telescope (KAGRA) in the Kamioka mine in Japan – a laser interferometer stretching across 3 kilometres. The NAOJ aims to connect it with the Laser Interferometer Gravitational-Wave Observatory (LIGO) facilities in the United States and the Virgo detector in Europe. The idea is to make the largest gravitational wave detector possible. We will have a further look at these types of detectors in Chapter 6.

European Observatories

In Chapter 3, we traced the history of selected European facilities that arose in the years before the 20th Century. It's useful to look at what is happening in Europe today. While many older observatories – particularly those in larger cities that are plagued by light pollution – have either closed, become museums or are mainly used for university purposes, there remains a large collection of major facilities on the Continent and in Britain. Not surprisingly, some of the largest research facilities are now located well away from city lights, in fairly remote areas.

The Roque de los Muchachos Observatory on La Palma in the Canary Islands is a good example. It operates optical, solar, infrared, and gamma-ray facilities. It's under the administration of the Instituto de Astrofisica de Canarias, under the umbrella of the European Northern Observatory, which also operates observatories on Tenerife. The observatory was first founded in 1984, when the Isaac Newton Telescope moved there from its prior location at Herstmonceux

Castle in England. Over the past few decades, the observatory has expanded to include the 4.2-metre (13.7-foot) William Herschel Telescope, the 3.58-metre (11.7-foot) Galileo National Telescope, and the 10.4-metre (34-foot) segmented-mirror Gran Telescopio Canarias. It remains both a productive facility, as well as a tourist attraction set among the rocky hills of the island of La Palma. As with facilities in the south-west USA, Hawaii's Mauna Kea and other remote locations, the weather on La Palma is very dry and allows for near-perfect observing conditions.

The Teide Observatory, located at 2,390 metres (7,841 feet) above sea level on Spain's Island of Tenerife, contains four solar telescopes, eight optical telescopes, and four radio telescopes. Built in the 1960s, it was the first of the major international observatories and it's operated by the Instituto Astrofisica de Canarias. Over the years, institutions from around the world built telescopes there to take advantage of the same dry climate that enabled the later construction of the facilities at La Palma. There are currently four solar telescopes, as well as installations for robotic observing, infrared astronomy, and visible-light observations at the Teide Observatory. The site also hosts several radio telescopes, as well as a visitors' centre where tourists can learn more about the observatory and its history.

Spain boasts the Calar Alto Observatory, located in Almeria, Andalusia. This optical facility, which was owned by the Max Planck Institute in Heidelberg until 2019, is operated by the Council of Andalusia and shares its management with the Spanish National Research Council. The observatory has four principal telescopes on site, ranging in size from 3.5 metres (11 feet) to 80 cm (31 inches). Advanced astrophysics projects include the Calar Alto Legacy Integral Field Area Survey, which mapped 600 galaxies over several years. The aim of the project is to understand the chemical evolution of galaxies, determining galaxy masses, and studies of structure and star formation within galaxies.

One of the radio astronomy facilities spread throughout the United Kingdom and Europe is the Lovell Telescope at Jodrell Bank Observatory, a fully steerable dish measuring 76.2 metres (250 feet) across. It's named after Sir Bernard Lovell, the English radio astronomer and physicist who was the first director at Jodrell Bank Observatory, which is part of the Jodrell Bank Centre for Astrophysics at the University of Manchester. The telescope has been involved in some of the most pivotal astronomy research since its commissioning, including in-depth observations of hydrogen clouds in our own galaxy as well as in inter-galactic space, detecting radio signals from neutron star mergers, and studies of emissions from objects in our own solar system. Jodrell Bank is also headquarters for the Square Kilometre Array in Australia and South Africa.

In Europe, the largest radio telescope is in Effelsberg, Germany, southwest of Bonn. This 100-metre (328-foot) fully steerable dish was built by the Max Planck Institute for Radio Astronomy and began observations in 1972. It can operate independently or be linked into an array, as needed. Since its commissioning, the observatory has been used for studies of solar system objects, gravitational lensing at radio wavelengths, magnetic fields in galaxies, and measurements of energetic objects such as the pulsar at the heart of the Crab Nebula.

One of the more interesting observatories in Europe is the Low Frequency Array (LOFAR). Most of radio receivers are spread out across the Netherlands. There are also antennas in Germany: at Effelsberg, Garching bei Munchen, Tautenburg, Potsdam and Jülich. One antenna is installed in the United Kingdom, at Chilbolton Observatory; while others are in Nançay, in France; Onsala Space Observatory, in Sweden; at three sites in Poland; at Birr Castle, in Ireland; and in Italy. LOFAR's aim is to map the universe at radio frequencies between 10–240 MHz. This wide-spread array gives astronomers an observing 'eye' of just over 300,000 square metres (186 square miles). The observatory is administered by ASTRON, the Netherlands research organisation. Although it operates in a radio-loud environment in Europe, LOFAR scientists have devised ways to filter out noise to get the science they want.

As with other research institutions around the globe, many European science organisations and universities have invested and/ or partnered in new facilities in more remote locations. Since most of these facilities are located south of the equator, our next step in the virtual observatory tour heads 'down under'.

Observatories South of the Equator

For many years, the bulk of the world's observatories were located mostly in the Northern Hemisphere. This, despite the fact that some of the most interesting research objects lie in southern hemisphere skies. There were, of course, a few large research stations south of the equator, in Australia and South Africa, and a very few in South America. Until the early 20th century, observations of southern skies were largely expedition-based, mounted by teams of astronomers from the US and Europe. This changed by mid-century, and astronomers began looking south for clear skies and calmer atmospheric conditions.

The rugged terrain of Chile's mountains and high-altitude deserts has emerged over the past few decades as a pre-eminent place for observatories. These regions are particularly useful for both optical and infrared observations, similar to the environment in Hawaii and on the Canary Islands. In addition, the plains of Chajnantor, high in the Andes, are one of the most radio-quiet places on the planet. Those attributes have attracted astronomers from around the world to observe at world-class institutions run by a number of university and governmental consortia.

Astronomy at Cerro Tololo

In the 1960s, astronomers connected to the Associated Universities for Research in Astronomy in the US began looking over sites for a new observatory in Chile. It eventually became the Cerro Tololo Inter-American Observatory (CTIO). Today it comprises a collection of telescopes located about 80 kilometres outside of La Serena, Chile, at an altitude of 2,200 metres (7,217 feet). Over the years, the telescopes at Cerro Tololo have been used to make major discoveries, ranging from dwarf planets around other stars to the characteristics of distant quasars and black holes. Currently the operations are funded by the US National Science Foundation. The Gemini Observatory South telescope is located nearby.

The main telescopes at CTIO are the 4-metre (13-foot) Victor M. Blanco telescope and the Southern Astrophysical Research telescope. There is also a collection of instruments and cameras that allow astronomers to conduct imaging and spectroscopy at visible and infrared wavelengths. One of the more intriguing projects done at

CTIO is the Dark Energy Survey. The study, which ended in January 2019, used a specialised 512-metapixel camera mounted on the 4-metre telescope. The study serves a collaboration of more than 400 scientists from around the world. The Dark Energy Camera (DECam) was used to do a survey of 300 million galaxies spread out across 5,000 square degrees of sky. It obtained information about each galaxy in an effort to learn more about this still-mysterious force affecting the acceleration of the expanding universe. DECam's data will be used to design new surveys and probes of the universe and, perhaps someday, they will give a much better idea of what dark energy is. In the meantime, data from the camera has been used to study variable stars, streams of stars that have been cannibalised into the Milky Way, and it has found evidence for more dwarf galaxies orbiting along with our galaxy.

European Southern Observatory (ESO)
CTIO is not the only set of observatories in Chile. It is joined by the European Southern Observatory, run by a 16-nation consortium of research facilities at governments and universities. Member states are Austria, Belgium, Czech Republic, Denmark, Finland, France, Germany, Ireland, Italy, the Netherlands, Poland, Portugal, Spain, Sweden, Switzerland, and the United Kingdom. The organisation has telescopes at La Silla, Paranal, and Llano de Chajnantor, and has built some of the largest and most technologically advanced telescopes in the world.

Astronomy at La Silla
The La Silla Observatory is located on the edge of the Atacama Desert, 600 km (372 miles) north of Santiago, and lying at 2,400 metres (7,874 feet). This location enjoys mostly dry conditions and is well away from sources of light pollution. This gives it one of the darkest night skies on the planet. La Silla is home to the 3.58-metre (11-foot) New Technology Telescope, as well as the ESO 3.6-metre telescope, which is outfitted with a planet-hunting spectrograph called the High Accuracy Radial Velocity Planet Searcher.

The La Silla site began development and construction in 1964 and was dedicated on 25 March 1969. The ESO 1.5-metre (4.9-foot) and 1-metre (3.2-foot) telescopes were built first. Less than a decade later, in 1976, the ESO 3.6-metre (11-foot) telescope was dedicated and began operations on the mountain. A 2.2-metre (7.2-foot) telescope was built in 1984, followed by the New Technology Telescope in 1989. In 1984, the 2.2-metre Max Planck-ESO telescope began operations, while in March 1989, the 3.5-metre (11.4-foot) New Technology Telescope (NTT) saw first light.

Today, La Silla is also home to the Swedish ESO Submillimetre Telescope, a 1-metre (3.2-foot) telescope operated by the Marseille Observatory, and a 1.2-metre (3.9-foot) Euler telescope from Geneva. Currently the observatories at La Silla are involved in cutting-edge observations. The La Silla-QUEST survey, for example, is searching the solar system for distant objects. A robotic exoplanet survey, called TRAPPIST (the TRAnsiting Planets and Planetesimals Small Telescope-South) is searching both planets around other worlds, but also looks for comets. Its first discovery, called TRAPPIST-1, turned out to be a planet system of seven near-Earth-like planets orbiting an ultra cool dwarf star.

Observing from Cerro Paranal
The European Southern Observatory (ESO) maintains a second observing reserve on Cerro Parana in the Atacama Desert. The mountain rises up to 2,636 metres (8,648 feet) above sea level and, like its sister high-altitude sites, offers ideal observing conditions. There are seven major telescope at Paranal, including the Very Large Telescope, which is the main facility on the mountain. The rest of the facilities contain survey instruments that scan and image the sky in optical and infrared light.

The Very Large Telescope: One Facility, Four Telescopes
The Very Large Telescope is ESO's crown jewel of observing platforms. Set high above the surrounding mountains and valleys, this four-telescope facility boasts four individual telescopes, complete with auxiliary telescopes and complete suites of instruments for each facility. The heart of each of the four is an 8.2-metre (26-foot) mirror. The telescopes can be used individually for observations, or ganged together to get very high resolution views of distant objects in both optical and infrared wavelengths of light.

Each of the telescopes has a unique name, based on the indigenous Mapuche language: Antu, Kueyen, Melipal, and Yepun. Since their commissioning began in 1999, the four telescopes have been in heavy use, and ESO claims they are the most productive observatories in the world, with an average of a new paper about science performed at the facility published each day. Over the years, the telescopes have delivered unique results, ranging from images of distant exoplanets, motions of stars around the black hole at the heart of the Milky Way, the detection of carbon monoxide molecules in a galaxy some 11 billion light-years away, it has peered into star birth regions in the Orion Nebula, and watched as a sun-like star sends its outer atmosphere to space in a gaseous shell of debris.

The Plains of Chajnantor

Some of the world's most cutting-edge radio astronomy happens in the 5,100-metre-high (16,732-foot-altitude) Llano de Chajnantor, a plateau in the Atacama Desert. It's the world's driest desert, ringed with rock formations, with hot summer temperatures and freezing cold winters. Yet this place, where the atmosphere is so thin that it literally takes one's breath away, is home to one of the most modern and sensitive radio arrays in the world. The Atacama Large Millimetre Array (ALMA), which is administered by a consortium consisting of the European Southern Observatory, the US National Radio Astronomy Observatory, and the National Astronomical Observatory of Japan, sprawls across up to 16 kilometres of this harsh landscape. The array consists of fifty 12-metre (39.3-foot) diameter antennas sensitive to signals in the submillimetre range of the electromagnetic spectrum. Like its sibling VLA in New Mexico, ALMA dishes can be grouped together in a tight configuration or spread out to create a giant 'virtual' dish. Technical staff move the telescopes using specialised transports. The main ALMA dish collection is complemented by the Atacama Compact Array, a set of sixteen dishes that can be added to science observations to gain better resolution. The complete ALMA array began operations in 2013 after a ten-year planning and construction process. It cost approximately $1.4 billion to build.

The Atacama Large Millimeter Array (ALMA) in Chile. This pictures shows a part of the telescope array in operation. The array sits on the Chajnantor plains and has up to sixty-six high-precision antennas working together at millimeter and submillimeter wavelengths. Courtesy of NRAO/AUI/NSF, Jeff Hellerman.

ALMA Science

In its first years of operation, ALMA has observed objects and events ranging from colliding galaxies and planet formation around distant stars. It has been used to follow eruptions of a volcano on the Jovian moon Io, and tracked organic molecules in material orbiting a young star. In one case, astronomers using ALMA found a veritable soup of organic molecules in the region around a very young, active star called V883 Orionis. It's surrounded by a disk of gas, dust, and ice, which may form planets in the distant future. Since the star is young, it's prone to outbursts, and one of those outbursts heated up the disk, which released molecules of methanol, acetone, and other organic compounds that had been locked away in the disk's ices. Astronomers want to know how similar this is to events in our own solar system some 4.5 billion years ago, and what such events can tell us about the formation of our own retinue of planets, asteroids, moons, rings, and comets.

ALMA and its sister radio observatories around the world can sense frequencies emitted by molecules in space which aren't detectable by other types of observatories. The waves it detects are not scattered or reflected by interplanetary dust, and thus can 'get through' clouds of material that obscure other wavelengths and frequencies.

In 2018, the array was part of the Event Horizon Telescope, which studied the black hole at the heart of the elliptical galaxy M87. Its data were combined with observations from nine other radio telescope facilities around the world to create the first-ever image of the event horizon around the black hole. The ALMA facility allows astronomers to map the distribution of gas and dust in our own galaxy, as well as in other nearby galaxies.

Australian Observatories

Australia is in the vanguard of world astronomy, offering scientists a range of optical, infrared, survey, solar and radio telescopes in astronomy reserves. Visiting them all is a continent-wide endeavour, with facilities in New South Wales as well as Canberra, Melbourne, the Outback, and Perth. Yet, these far-flung institutions are providing astronomers with unprecedented access to the sky, our galaxy, and the universe beyond.

Mount Stromlo (ACT) Rises Again

On 18 January 2003, the oldest astronomical institution in Australia's Capital Territory, the Mount Stromlo Observatory, was almost completely destroyed by a wildfire. For many Australians (and indeed, for astronomers on Mount Wilson and other mountaintop observatories in the world), such wildfires are a fact of life. Mount Stromlo survived the fire with few facilities intact, and it seemed like a fatal blow to the Australian astronomy community. Not only were the telescopes destroyed or damaged, including the Great Melbourne Reflector that had been moved to the mountain, but the library, workshops, the director's home and other residents, as well as the observatory's archives containing a sizeable part of the country's astronomical research history, were all heavily damaged or burnt to the ground. A cutting-edge instrument called Near Infrared

Integral Field Spectrometer, which was being built on the mountain for the Gemini Observatory, was also destroyed. Only one telescope survived. Still, the community persisted and, at the time of writing, more than a decade and a half after the devastation, the observatory has rebuilt.

Mount Stromlo's use as an astronomy site dates back to the early 1900s, when it hosted the Commonwealth Solar Observatory with a focus on solar physics and atmospheric studies in the 1920s. After the Second World War and its use as a munitions area, astronomers began using the facilities for studies of stars and galaxies. The name changed to the Commonwealth Observatory. Today, it operates as the Mount Stromlo Observatory under the auspices of the Australian National Observatory Research School of Astronomy and Astrophysics. Beginning the 1990s, when Australia became one of the partner countries in the Gemini Observatory project, Mount Stromlo shifted some of its focus to building instrumentation. Mount Stromlo is currently at the forefront of instrument development with an Advanced Instrumentation Technology centre. It also operates a SkyMapper Survey telescope, and astronomers are working on new instruments and updates for the 2.3-metre (7.5-foot) telescope. In particular, the Gemini NIFS instrument was rebuilt, and the observatory also built the Gemini South adaptive optics imager, which is now deployed at Gemini in Chile. As part of the Australian National University, Mount Stromlo is also involved in the construction of the Giant Magellan Telescope, set for commissioning around 2020.

The observatory grounds are not far from the city lights of Canberra, the country's capital. Those same lights threatened the viability of the observatory long before the 2003 fire consumed nearly the entire facility. Starting in the 1960s, in an effort to sidestep the light pollution problem, the University established a second major observatory at Siding Spring, in the Warrumbungle Mountains. The joint facility took the name 'Mount Stromlo and Siding Spring Observatories'. The joint facilities now perform much of Australia's optical astronomy, with astronomers using the telescopes for stellar and galactic studies, the structure and evolution of planets, stars and galaxies, the origin and continued evolution of the universe as a whole, and studies of the interstellar medium.

Astronomy at Siding Spring

Mount Stromlo's partner facility, the Siding Spring Observatory at Coonabarabran, New South Wales, boasts the 3.9-metre (12.8-foot) Anglo-Australian Telescope (AAT), the 2-metre (6.5-foot) Faulkes Telescope, the 2.3-metre (7.5-foot) Advanced Technology Telescope, plus a dozen other facilities and instruments. Astronomers from around the world use these telescopes for a variety of research programmes and automated sky surveys.

One of the most fascinating projects undertaken at the observatory was the 2DF Galaxy Redshift Survey. This was a study of galaxies in a 1500-square-degree area of sky, conducted using a specialised instrument installed on the 3.9-metre telescope. It measured the redshifts of more than 220,000 galaxies during the period from 1995 to 2002. The result was a three-dimensional map of that region of the sky, looking out across more than three billion light-years. The 2DF survey was the first map of what's called the large-scale structure of the local universe. Large-scale structure refers to the distribution of galaxies in a web-like arrangement of galactic sheets, filaments, and voids. Understanding this structure in the local as well as the distant universe gives astronomers a way to measure the evolution of the universe and also get a handle on the distribution of dark matter.

Following the 2DF survey, astronomers used the same telescope to do similar wide-scale observations of quasars, called the 2DF QSO Redshift Survey. Then, in 2001, the 6DF survey began, with a team of astronomers using a new instrument to look at more than a hundred thousand galaxies, their redshifts and motions. Along with the Sloan Digital Sky Survey (which also studied quasars in its field of view), these redshift observations have given astronomers new insight into the how galaxies cluster and move in an expanding universe.

Radio Astronomy in Australia

Australia is also home to the famous Parkes Observatory 64-metre (209-foot) radio telescope, which sprang to fame as part of the network which received messages from the Moon during the Apollo 11 landing. Located in the small town of Parkes in New South Wales, the telescope dominates what looks like quite a pastoral landscape. Parks has been in operation since 1961, and is part of the Australia Telescope National Facility, a wide-ranging network of radio telescopes that also includes the Australia Telescope Compact Array at Narrabri, and a single 22-metre (72-foot) dish near Coonabarabran called the Mopra Radio Telescope. Aside from its fame as part of the Apollo 11 mission, Parkes has also been involved in tracking various spacecraft, including two Mariner missions, the Voyager missions during part of their trek, the Giotto spacecraft to Comet Halley, the Galileo mission to Jupiter, and the Cassini-Huygens mission to Saturn.

The western end of the country is home to an unusual installation of very low-frequency telescopes which make up the Murchison Wide-field Array. It is located near Boolardy in the western Outback, and is made up of hundreds of 'tiles' containing dipole radio antennas. The entire array is sensitive to a frequency range of 80 to 300 MHz. This is also a frequency range that is easily disrupted by radio and television broadcasts, as well as electronics from large vehicles. The remote location of the MWA was chosen as one of the most radio-quiet places on Earth, and it's now being operated by a consortium of twenty university and governmental partners from around the world.

Why do astronomy at the low end of the radio spectrum when most radio astronomy is done at higher frequencies? It turns out some interesting events are much 'brighter' at low frequencies. Thus, MWA was designed to observe a 30-degree-wide field of view of the sky, including the plane of the Milky Way galaxy to study naturally occurring low-frequency signals from such objects as HII regions in the galaxy. These are clouds of molecular hydrogen (hence the name HII) where stars may form.

In addition, MWA can track space debris, make observations of halo stars and structures in ancient galaxy clusters, track radio sources in the sky, and detect of transient radio sources. During its early phase of commissioning, the array did a survey of 300,000 radio sources outside our own galaxy. It also was used to track plasma tubes in Earth's ionosphere. Scientists hope to use the array to study signals dating back to the epoch of re-ionisation, a period some 300 to 400 million years after the Big Bang, when the infant universe had cooled enough to permit light to travel freely.

The MWA is also a precursor instrument for the nearby Square Kilometre Array, which will be built in both Australia and South Africa. It is a planned array of radio telescopes that would yield a signal collecting area of one square kilometre. The Australian pathfinder SKA is located not far from MWA. We will read more about SKA in Chapter 7.

Astronomy in South Africa

As with so many of the observatories south of the equator, astronomy in South Africa began when sailing ships from the Northern Hemisphere began venturing south in search of trade and other commodities. Astronomers soon followed, looking for good skies and chasing specific events such as eclipses, comets, and transits. The first observatory in South Africa was built in the 1820s, in what is now known as the suburb of Observatory in Cape Town. Today that building is the headquarters of the South African Astronomical

The telescopes of the South African Astronomical Observatory at Sutherland in the Karoo, a semi-desert area. Credit P. Schella, used by permission, Wikimedia Commons.

Observatory (SAAO). The actual telescopes it operates are located well away from the light pollution of the city, about 300 kilometres (186 miles) away in Sutherland. The SAAO maintains the South African Large Telescope (SALT), an 11-metre (36-foot) telescope, which is funded and used by a consortium of astronomers and institutions in Germany, India, New Zealand, Poland, the UK and US in addition to South Africa. Along with other facilities, SALT is involved in the search for dark matter, and studies lensing transient events in the sky, variable stars, and much more.

The SAAO Sutherland site also hosts telescopes for other institutions, such as the MeerLICHT (which looks at optical counterparts of events observed by the MeerKAT radio array). Nearby is the Birmingham Solar Oscillations Network Sutherland station (BiSON), which is part of a network of solar telescopes that track 5-minute-long oscillations of the Sun. Korea has a microlensing survey telescope called the Korea Microlensing Telescope Network (KMTNet), and the Los Cumbres Observatory private robotic network has a 1-metre telescope installed there, as well. (We will read more about Los Cumbres in Chapter 8). To hunt for extrasolar planets, a group of institutions in Poland is launching a global network of robotic telescopes. The project, called Solaris, has installed two telescopes at SAAO.

Antarctica

It doesn't get much farther south than Antarctica when it comes to doing astronomy. Visitors to this South Polar Region tell tales of penguins and ice sheets, but astronomers see it as a perfect place to get completely away from all the light and radio frequency pollution that plagues many other sites in the world. Antarctica combines high altitude with dry climate, and, for at least part of the year, the continent is completely dark during winter. There have been several observation stations set up in the region, and a number of balloon-borne telescopes have operated there, as well. (Those will be discussed in Chapter 6.)

Antarctica is home to the South Pole Telescope (SPT), which is located at the Amundsen-Scott Station and was commissioned in 2007. It's run by the University of Chicago and a group of collaborators including the University of California at Berkeley,

Case Western Reserve University, the Smithsonian Astrophysical Observatory, McGill University in Canada, the University of Colorado at Boulder, the Argonne National Laboratory, and the University of California-Davis.

The heart of the SPT is a 10-metre (32-foot) primary mirror, with secondary mirrors and an array of detectors to collect the incoming millimetre-wavelength light. The main goal of the SPT is to study the cosmic microwave background, and also looks for evidence of dark energy that could be shaping galaxy clusters. Dark energy is a so-far unknown form of energy which appears to be pushing everything in the universe apart. This telescope was also used as part of the Event Horizon Telescope to capture the first-ever radio 'image' of the event horizon around the supermassive black hole in the distant elliptical galaxy M87.

There have been other telescopes installed and used at the South Pole, including the robotic High Elevation Antarctic Terahertz Telescope (HEAT). It has been operated during each local winter by the University of Arizona and the University of New South Wales in Australia. It was built to do spectroscopic mapping of the plane of the Milky Way Galaxy at far-infrared wavelengths.

Moving on Up

As we've seen from this tour of selected observatories, astronomy is a world-wide endeavour. Astronomers continually work to get better views, build bigger and better telescopes, all to learn more about the universe. In recent decades, the acuity and data-gathering capabilities of computer-enabled telescopes, and their ingenious instrument packages, has revealed the innermost workings of planets, stars, and galaxies. Yet, there are still regimes which cannot be studied from Earth, no matter how good facilities are or how perfect the seeing conditions can be. To extend our senses beyond our planet, we have to send our telescopes away from home, to orbit well above the atmosphere. Space astronomy has enabled incredible discoveries and, as we shall see in the next chapter, it continues to widen our view of the cosmos.

5

EYES IN THE SKY: SPACE-BASED OBSERVATORIES

Hardly a week goes by without images of celestial objects captured by the world's fleet of space telescopes appearing in the media. They are so routine as to become a fact of life: a pretty nebula, a glittering star cluster, an artist's conception of a distant exoplanet crossing in front of its star, to views of ancient galaxies. That almost simple-seeming availability of such imagery doesn't come easily or affordably. Yet, thanks to orbiting observatories, we can peer at the universe over the shoulders of astronomers – out to the surfaces of planets in our solar system, to planets around other stars, to the central regions of our own galaxy, and outwards to objects so far away their light took billions of years to reach us.

Why Go to Space?
What are these observatories and how did they come about? Doing astronomy from space is not a new idea, although it wasn't possible until the 1960s. It was, however, a huge technological leap from the kind of observations astronomers were used to doing with ground-based facilities. Think about what's been discussed in prior chapters: the development of telescopes and the environmental conditions they faced that were largely solved by enclosing them in buildings. Now, think about the rigors of space and how that complicates the process of building a telescope off-Earth. Any

spacecraft – whether it's carrying crew to a space station or delivering a telescope payload to orbit (whether around Earth or another world) – faces tremendous challenges to its very existence in space. The entire rocket and its cargo have to withstand the launch out of the gravitational well of our planet and then into the very low-gravity or micro-gravity environments where they will operate. Space presents extreme conditions: searing heat, freezing cold, and strong radiation blasts. In space, the telescope's mirrors must remain aligned, its hardened electronics have to perform their tasks consistently, the on-board computers must function properly, and the power supplies have to last the length of the nominal mission (and often, they last beyond that). It's a lot to ask of an instrument which will be performing precision measurements for years.

It's no small task to prepare a telescope for launch into space. Long before it is sent to orbit, astronomers and technical wizards plan and build their instruments. They have to make them fit exacting weight and mass standards, even as they prepare telescopes that have to withstand so much punishment in space. It takes teams of hundreds or thousands of experts to do this: the astronomers who know what science observations they want to perform, the technical experts to draft and assemble the craft, the agencies to fund the work, and the companies and institutions who launch the payloads. So why do it? Yes, it's hard, but astronomers and space agencies have been sending scopes to space for decades. The payoff: great science which helps us understand our universe. It's science not easily obtained from ground-based telescopes; in some cases, as we've seen in earlier chapters, some of it is science which can't be gotten from the ground *at all*.

A Brief History of Space Telescopes

Most readers are familiar with the long-running Hubble Space Telescope. It was launched in 1990 and its history of observations has been successful beyond its designers' wildest dreams. However, while it may be the avatar of orbiting telescope success, it is far from the first to be launched. Ideas for orbiting space observatories date back to the early decades of the 20th century, when rocket scientists such as the German expert Hermann

Oberth first dreamed of putting a similar type of observatory on an asteroid and sending regular supplies up to its orbiting 'night assistants' who would live aboard the telescope station for varying lengths of time. Before that time (and since), astronomers have also discussed lunar-based observatories, which were also a staple of science fiction. Actually, the idea of telescopes on the Moon is not so far-fetched. During its 14-day night, the lunar surface provides excellent views of the sky, and with a little planning, radio telescopes can be situated out of the din of Earth's radio frequency interference. Such observatories may well be built in the next few decades.

Oberth, and later on the astrophysicist Lyman Spitzer (who went on to promote the idea of the Hubble Space Telescope), knew that Earth's atmosphere posed a huge problem for observers – not just because of atmospheric aberration, but also because the atmosphere absorbs great swaths of the infrared and ultraviolet spectrum, as well as x-rays. To get above atmospheric interference, astronomers began launching telescopes on short suborbital hops aboard sounding rockets, beginning in 1945, just as the Second World War was winding down. Sounding rockets carry detectors and instruments into Earth's upper atmosphere, usually for short periods of time. The first ones were really sent to measure high altitude atmospheric characteristics. Robert H. Goddard, the US rocket genius, originally developed his rockets for this purpose. Sounding rockets were a staple of exploration throughout the 1950s, and helped bring about three new scientific disciplines: gamma-ray astronomy, x-ray astronomy, and ultraviolet astronomy. Mysterious objects suddenly came into view, such as Scorpius X-1. It turns out to be a very strong x-ray source and is a neutron star orbiting a smaller companion. Scorpius X-1 was discovered from an Aerobee sounding rocket in 1962.

The International Geophysical Year (IGY) was the impetus for sounding rocket research in the late 1950s. It began a world-wide scientific effort to a serious study of our planet and its interplanetary environment. The IGY began on 1 July 1957 and ended on 31 December 1958. Eleven scientific disciplines were included in the effort: auroral studies, cosmic ray studies, geomagnetism, gravity, ionospheric physics, longitude and latitude determinations, meteorology, oceanography, and solar activity. As it happened, the time also coincided with the peak of a solar cycle – that period of eleven years which goes from sunspot minimum to sunspot maximum and back again. It was a period of rocket launches, balloon studies (mentioned below), and – to the surprise of most of the world (including many Russians) – the launch of the first artificial satellite, Sputnik 1. Many of the Earth-observations studies (mentioned later in this chapter) got their start during this period. The IGY also brought about closer international cooperation in science, and led to some startling discoveries, among them the existence of the Van Allen Radiation Belts high above Earth.

Balloons in Astronomy

Sounding rockets weren't the only methods astronomers used to get instruments to extremely high altitudes. The Stratoscope I and II projects were launched starting in the late 1950s. The first was used until the 1970s, and took a 30.48-cm (12-inch) mirror and telescope to space. It was used to look at activity in the Sun's photosphere. The second one lofted a 91.4-cm (3-foot) telescope to the upper atmosphere and flew on missions starting in 1963. Its main targets were planetary atmospheres, aging stars, and distant galaxies. The main idea was to get these telescopes above the atmosphere for long-term studies.

In 1997, scientists interested in studying cosmic background radiation began a series of high-latitude balloon flights carrying the Balloon Observations of Millimetric Extragalactic Radiation and Geophysics (BOOMERanG) to gather data about this faint light that was first radiated a few hundred million years after the Big Bang. It made flights until 2003 and made significant discoveries.

Satellite Observatories Come into Their Own

NASA launched a series of eight Orbiting Solar Observatory satellites, beginning in 1962. Six Orbiting Geophysical Observatories were sent into orbit beginning in 1964, and were built to study Earth's atmosphere and magnetosphere and scan the region of space between Earth and the Moon. Eventually, astronomers wanted to look beyond the Sun, and that led to the development of four Orbiting Astronomical Observatories (OAO). NASA launched them, starting in 1966, and they provided the first high-resolution ultraviolet studies of astronomical objects.

OAO-1 carried ultraviolet, x-ray and gamma-ray sensitive instruments, but a power failure aborted the mission not long after launch. OAO-2, also called Stargazer, was a successful mission launched on 7 December 1968. It was equipped with 11 ultraviolet telescopes. It watched novae (outbursting stars) and also detected huge clouds of hydrogen molecules surrounding comets as they got close to the Sun. OAO-B carried a sizeable 97-cm (38-inch) telescope, but the mission failed and never made it to orbit. The last mission, OAO-3, known as Copernicus, was launched on 21 August 1972 and gathered high-resolution spectra of stars and made x-ray observations. The OAO series was the nearest precursors to the Hubble Space Telescope (which was also based on the later Keyhole reconnaissance design) and set the stage for the development of other orbiting observatories.

Since those early days, nearly a hundred space observatories have been planned, built, and sent to space by NASA, ESA, China, Japan and consortia of countries around the world. About twenty are in orbit right now, and more will come online in the next decades. The age of space-based astronomy is well underway. Some of the orbiting telescopes study the universe in very limited wavelengths, while others are multi-wavelength facilities. Some gather up cosmic ray particles, and eventually, a full array of gravitational-wave sensing telescopes will be on orbit.

If we count solar system probes as observatories (and some space experts do not), then we can add nearly 180 other explorers that are equipped to observe very specific targets. Not all the missions were successful but many were, and from them we have learned much about objects and processes in the solar system. Regardless of what they study, the deployment of orbiting and target-based telescopes and probes continues. New telescopes, such as the James Webb Space Telescope are coming online in the next decade and others are still in the planning stages.

Orbiting Telescopes

From its very earliest times, astronomy has been a science of measurement and positions, in addition to a search for cosmic understanding. The ancient astronomers who charted positions of celestial objects set those charting and mapping activities, and today's astronomers continue the work in the discipline of astrometry.

The Gaia space observatory is on an astrometric mission to measure the positions, distances, and motions of stars in our own galaxy with the highest possible precision. Launched by the European Space Agency in 2013, Gaia is gathering data that will enable astronomers to create an incredibly precise three-dimensional map of objects in the Milky Way Galaxy. The spacecraft, which (as of this writing) is about two-thirds of the way through its mission, does this in three ways: with the Astro instrument, which determines precise positions of stars between magnitudes 5.7 to 20 (stars much fainter than the Sun); with the photometric instrument, which makes measurements of an object's luminosity; and the radial-velocity spectrometer, which does spectral analysis of the light from objects and determines its velocity. Since its launch, Gaia has released two sets of observational data that includes positions and magnitudes for billions of stars, as well as colour data and velocity information. More data will be forthcoming in the 2020s.

Although checking the positions of stars in the Milky Way sounds like it could be fairly routine, the mission has turned up some quite interesting finds. For example, Gaia captured information about the Sculptor dwarf galaxy. This is a spheroid-shaped galaxy that actually is a satellite of the Milky Way. In November 2017, scientists released Gaia results showing this tiny galaxy follows a highly elliptical orbit with respect to the Milky Way. Currently it is quite close to our galaxy, and with this latest data, astronomers know it is making a close approach to the Milky Way at a distance of about 272,000 light-years.

It's been well known for years that the Milky Way is orbited by a flock of little galaxies, and that it's in the process of cannibalising some of them. Using Gaia measurements of stars, astronomers also found a new, very small and dim satellite galaxy called Antlia 2. Gaia has also uncovered some surprises inside our own galaxy. It turns out there is

Above: Stonehenge, located in Wiltshire, England, is thought to be a Neolithic monument built as a burial ground, with the layout of some of the stones appearing to line up with sunrise on the summer solstice and sunset of winter solstice. Image by Gareth Wiscombe via Wikimedia, CC-BY 2.0 licence.

Below left: A fresco painted by Giuseppe Bertini showing the astronomer Galileo Galilei showing the Doge of Venice how to use his telescope in 1609. Galileo built and sold versions of this telescope. Public domain image.

Below right: An engraving from Tycho Brahe's book *Astronomiae instauratae mechanica*, published in 1598. It shows a brass quadrant and other instruments installed at his observatory at Uraniborg. Public domain image.

Above left: The Jantar Mantar at Jaipur in India has sundials, meridians and other instruments to measure sunrises, sunsets, transits, and other celestial events. There is a similar construction in Delhi. Wikimedia Commons licence Attribution-Share Alike 3.0.

Above right: Astronomer Giovanni Domenico Cassini (known in France as Jean-Dominique Cassini) had a meridian line inlaid into the floor of the San Petronio Basilica in Bologna, Italy. Sunlight enters through an opening in the wall, which projects light onto the meridian. The position of the sunlight on the line allowed church elders to determine the date and use that information to calculate when Easter would fall each year. Courtesy of L. Cassinam, CC BY 2.5.

Left: The Royal Observatory at Greenwich, showing the time ball. First established in 1675, the building is no longer used as an observatory for research, but is a museum. Courtesy of Kjetil Bjørnsrud, used by permission CC BY-SA 3.0.

Below: The Royal Observatory complex, established in Edinburgh, Scotland, was completed in 1834, and is still in use today. It also has facilities in the southern hemisphere. Public domain image.

Right: The Great Equatorial Telescope was commissioned for use in 1885, then moved to Herstmonceux in 1957. It is now back at the Greenwich site, refurbished and modernised with computer controls and used for educational programs. Courtesy of Geni@Wikimedia Commons.

Below left: Lick Observatory, with the Shane 3-metre (120-inch) telescope dome, center, and the Automated Planet Finder dome on the right. Courtesy of The Tahoe Guy; Creative Commons 2.0, Generic licence.

Below right: The Great Lick Refractor is 14 metres (57 feet) long and has a 91-cm (36-inch) aperture. It was installed in 1886 and is still in use today. Courtesy of Michael@Wikimedia. CC BY 2.0 usage licence.

Above left: An aerial view of Mount Wilson, located north of Los Angeles, California, in the San Gabriel Mountains. Courtesy of Doc Searls, CC BY 2.0 licence.

Above right: The 2.54-metre (100-inch) Hooker Telescope at Mount Wilson. It was completed in 1919 and is still in use today. Courtesy of Ken Spencer, used by permission, Creative Commons Attribution-Share Alike 3.0 licence.

The Clark Telescope Dome at Lowell Observatory in Flagstaff, Arizona. It houses the Clark Refractor, the original telescope purchased for use at the observatory by Percival Lowell. Courtesy of Pretzelpaws, Wikimedia Commons.

The Clark Refractor at Lowell. This telescope has been refurbished and has National Historic Landmark Designation. Percival Lowell used this telescope to observe Mars and Venus, and it has been used for more than a century for further research. Courtesy of Littlefield Energy, CC BY-SA 4.0.

The 5.1-metre (200-inch) Hale Telescope in California. For 45 years, it was the world's largest telescope and is still in heavy use by astronomers. Courtesy of Caltech/Palomar Observatory.

Griffith Observatory in Los Angeles, California. Built as a gift to the city from entrepreneur Griffith J. Griffith, it houses an observatory with both night-time and solar telescopes, plus a planetarium, and an extensive set of astronomy exhibits. It first opened in 1935, closed in 2002 for renovations, and re-opened in 2006. Courtesy Griffith Observatory.

Above left: The Sommers-Bausch Observatory on the campus of the University of Colorado. It opened in 1953 and is used today for public nights as well as for student and faculty observing projects. Courtesy M.E.C.U. CC BY-SA 3.0.

Above right: Two of the Kitt Peak National Observatory telescopes near Tucson, Arizona. On the left of the image is the Bok 2.3-metre (90-inch) telescope and on the right, the Mayall 4-metre (13-foot) telescope. Courtesy of Monte Best, CC BY 3.0.

Right: The Mayall 4-metre (13-foot) telescope at Kitt Peak National Observatory. Courtesy of Mark Hanna/ NOAO/AURA/NSF.

An aerial view of Siding Spring Observatory, near Coonabarabran, New South Wales, Australia. Courtesy of CC BY SA 4.0. S.S. Pete.

The observatory domes at the Cerro-Tololo Inter-American Observatory in Chile. Courtesy of T. Abbott and NOAO/AURA/NSF.

A panoramic view of the observatories on Mauna Kea, on the Big Island of Hawaii in 2019. There are thirteen facilities at the summit. Courtesy Frank Ravizza, Creative Commons Share-Alike 4.0.

A panorama of the ALMA Observatory antenna array with the Milky Way arching overhead. Courtesy of ESO/B. Tafreshi (twanight.org).

Gemini North telescope using its adaptive optics system to create a laser guide star. Courtesy Gemini Observatory.

Right: Sunset panorama shot from inside Gemini North, with all the vents open to help cool the telescope's mirror. Courtesy: Gemini Observatory.

Below: A time-lapse image showing star trails over Gemini North. Courtesy: Gemini Observatory.

Above left: The James Clerk Maxwell telescope, center, on Mauna Kea. The smaller dome at lower left is the CalTech Submillimeter Observatory; the buildings to the right of JCMT are the Smithsonian Submillimeter Array. Courtesy of A. Woodcraft. Public domain.

Above right: A 2009 aerial view of the Very Large Telescope Array at the European Southern Observatory at Paranal, Chile. The VLT is in the center; in the distance lies the Visible and Infrared Survey Telescope for Astronomy. Courtesy of J.L. Dauvergne & G. Hüdepohl (atacamaphoto.com)/ESO.

Left: The Very Large Telescope silhouetted at moonrise. Courtesy of ESO.

Below: The VLT Auxiliary Telescope shown silhouetted against the nighttime sky. The two bright objects in the sky in the center are the Large and Small Magellanic Clouds; this image shows the richness of the southern hemisphere sky. Courtesy of Y. Beletsky (LCO)/ESO.

Right: The Very Large Telescope array at night. Courtesy of Serge Brunier/ESO.

Below: Astronomers enjoy at sunset view of Paranal, Chile, with the Moon and Venus over one of the auxiliary telescopes for the Very Large Telescope Array. Courtesy of ESO/Y. Beletsky.

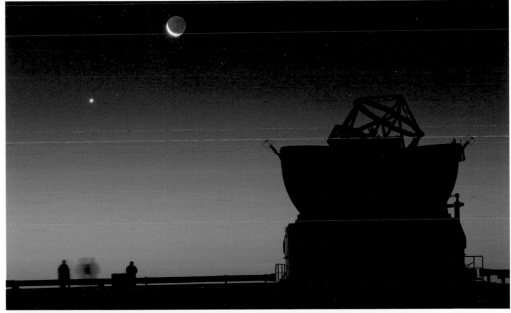

A 2001 view of the Isaac Newton Group of telescopes at Roque de Los Muchachos peak on La Palma in the Canary Islands. Courtesy of Nik Szymanek.

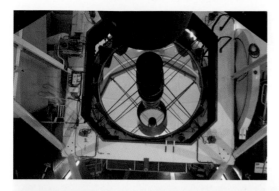

Left: The mirror of the William Herschel Telescope at La Palma. Courtesy of Nik Szymanek.

Below: The Milky Way Galaxy seen over the Karl G. Jansky Very Large Array west of Socorro, New Mexico. Courtesy: NRAO/ Jeffrey Hellerman.

A night-time view of the Allen Telescope Array at Hat Creek Observatory in California. This array is searching for possible signals from intelligent civilizations elsewhere in the galaxy. Courtesy SRI International.

A tile of low-frequency receivers being tested at the Murchison Widefield Array in Western Australia. Courtesy: Natasha Hurley Walker, Creative Commons Attribution-Share Alike 3.0 licence.

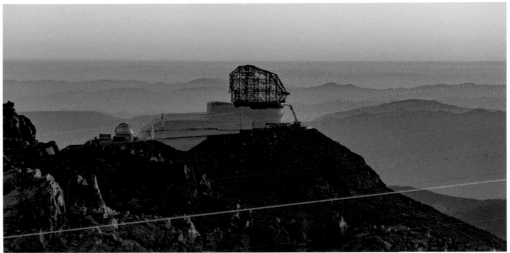

Above: The Large Synoptic Survey Telescope at sunset, as viewed from the Gemini South Telescope. The LSST is currently under construction. Courtesy of LSST Project/NSF/AURA.

Right: The LSST Primary/Tertiary Mirror (M1M3) arrived in the port of Coquimbo on 7 May 2019, was transported to the LSST summit facility building over the next several days and arrived on the summit on 11 May 11 2019. Courtesy of LSST Project/NSF/AURA.

Above left: Optical module in the lab showing how the module is supported by two parallel ropes for deployment as part of the KM3Net Neutrino detector. Courtesy KM3Net.

Above right: A demonstration of the IceCube neutrino detector array in use at the South Pole. Courtesy: IceCube/NSF.

The IceCube Laboratory at the Amundsen-Scott South Pole Station, in Antarctica, hosts the computers collecting raw data. Due to satellite bandwidth allocations, the first level of reconstruction and event filtering happens in near real time in this lab. Only events selected as interesting for physics studies are sent to UW–Madison, where they are prepared for, used by, any member of the IceCube Collaboration. Courtesy of Felipe Pedreros, IceCube/NSF.

Above and right:
The LIGO consortium
gravitational wave
detectors at Livingston,
Louisiana, and Hanford,
Washington. Courtesy
Caltech_MIT LIGO lab.

Below: A night-time
shot at the Kielder Star
Camp, held twice yearly
in England. Courtesy
Stuart Atkinson.

Amateur astronomer and astrophotographer Richard Wright with one of his test telescopes. Courtesy of Richard Wright.

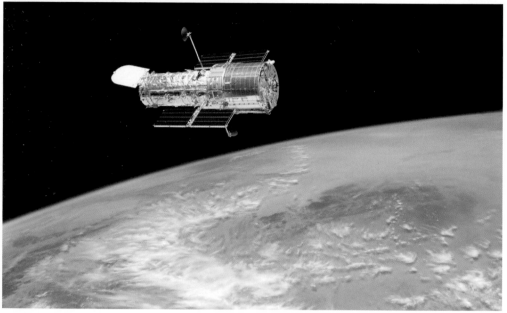

Above: Hubble Space Telescope just after it was released from the shuttle after a servicing mission in 2009. Courtesy NASA/ESA/STScI

Left: A 'selfie' taken by the Mars Curiosity Lander in May 2019. The spacecraft has been roving the surface of Mars since 6 August 2012. Courtesy: NASA/JPL-Caltech/MSSS.

An artist's conception of the Chandra X-ray Observatory during observations. Courtesy: NASA/CXC.

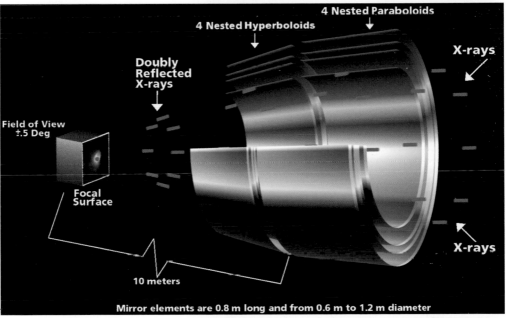

4 Nested Paraboloids

4 Nested Hyperboloids

Doubly Reflected X-rays

X-rays

Field of View ±.5 Deg

Focal Surface

X-rays

10 meters

Mirror elements are 0.8 m long and from 0.6 m to 1.2 m diameter

A diagram showing how x-rays are richocheted along Chandra's barrel-shaped mirrors toward detectors. Courtesy NASA/CXC.

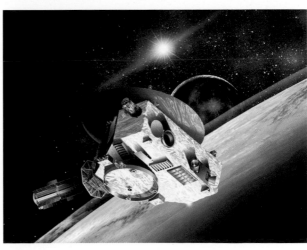

An artist's concept of the New Horizons mission as it flew by Pluto and Charon in 2015. Courtesy of Johns Hopkins University Applied Physics Laboratory/Southwest Research Institute (JHUAPL/SwRI).

The Gemini North, in the foreground, and the Canada-France-Hawaii Telescope domes in the pre-dawn twilight with the Milky Way arching overhead. Courtesy: Gemini Observatory.

a whole population of what are called hypervelocity stars. These are moving at upwards of 400 kilometres (~250 miles) per second. Some are leaving the Milky Way, while others appear to be headed toward our galaxy. That raises a lot of questions about them. Do they come from other galaxies? What kicked up the speed on those which appear to be leaving the Milky Way? It's possible that both the exiting and incoming stars are actually on very eccentric orbits through the halo of our galaxy.

Gaia data sets are also uncovering surprising new information about white dwarfs (which are ancient sunlike stars). It turns out these old stellar remnants can eventually turn into solid spheres as they cool. Other Gaia observations are changing our understanding of how star clusters form.

The Search for Distant Worlds

Of specific interest to astronomers are the search for and study of exoplanets (worlds orbiting other stars) and the continued search for the earliest objects to form in the infant universe. One of the most intriguing questions that astronomers can ask about the universe is, 'are there planets like Earth out there?' The search for exoplanets, as discussed in Chapter 1, has been a reality since the first one was discovered (using ground-based techniques) in 1995. Space-based observatories can play a big role teasing these faint objects out of the bright light of their parent stars.

Kepler

The Kepler Space Telescope did its planet-searching programme for more than nine years, from spring 2009 to late 2018. It was launched by NASA to look for Earth-type and Earth-size planets in the nearby part of our galaxy, within a few thousand light-years of the Sun. It did this with a 0.95-metre (3.1-foot) telescope at its heart, and transmitted data back to Earth for analysis. Once a planet candidate was identified, it was marked for further observations with other telescopes. Kepler used a technique called the 'transit method' to survey stars in the Milky Way in the direction of the constellations Cygnus, Lyra and Draco. Essentially, it looked at stars in a range of light encompassing all the visible and a little infrared, and watched as planets passing front of their stars (from our point of view) 'dimmed' the light. Using this method, the Kepler telescope found nearly 4,000 planetary candidates, and just over 1,000 of those have been confirmed as actual worlds via follow-up observations. Kepler only recently 'retired' when the fuel for its reaction wheels (gyroscopes, which keep it stable and pointed correctly) ran out.

TESS

A more recent entry in the planet-hunting race is the Transiting Exoplanet Survey Satellite (TESS). It was launched in 2018 and is actively searching for exoplanets. Unlike its predecessor, the now-defunct Kepler Space Telescope, TESS is searching the whole sky, whereas Kepler was focused on a relatively small part of the sky in the direction of the constellation Cygnus, the Swan. It uses light-sensitive devices to study flickers in brightness as planets 'transited' across the face of their stars, through four wide-angle view telescopes and CCD detectors.

The TESS cameras will focus on nearby stars that are type G, K, and M. Ultimately, the survey will look at half a billion stars, and will include some very close red dwarfs. Based on Kepler's findings, the TESS cameras should find tens of thousands of planets, including several hundred to a thousand Earth-sized and slightly larger worlds. One of its early results, announced in May 2019, the telescope spotted at least three super-Earths orbiting a bright star called HR 858. The new planets are (for now) called HR 858 b, c, and d. They have close, fast orbits around their star, and follow-up observations will clarify their orbital characteristics.

Since stars also vary in their brightnesses due to internal conditions and not necessarily due to planets 'dimming' them down slightly, TESS is also being used to study the asteroseismology of stars. It's the science that deduces the internal structure of stars based on their spectra and brightness changes. Interestingly, different 'oscillations' give clues to what's happening at different stellar depths.

A Closer Look at Hubble Space Telescope

Since we began this chapter with a mention of Hubble Space Telescope (HST), it's worth taking a closer look at this famous observatory and what makes it tick. HST orbits Earth at an altitude of about 547 kilometres (339 miles) and makes the trip around once every 97 minutes. It has been in orbit since 26 April 1990, the first of a group of space telescopes dubbed 'The Great Observatories' to be launched. It was built at a cost of $1.5 billion dollars and is managed by NASA. The European Space Agency is also a partner in the telescope and contributed some funding and a first-generation instrument. It also maintains a small staff presence at the Space Telescope Science Institute.

HST is outfitted with a 2.3-metre (7.5-foot) mirror, various scientific instruments and cameras. Astronauts performed several instrument 'swap outs' during servicing missions. Currently, it contains the Wide Field Camera 3, the Cosmic Origins Spectrograph (sensitive to ultraviolet light), the Advanced Camera for Surveys, the Space Telescope Imaging Spectrograph (sensitive to visible, some ultraviolet and near-infrared light), and the Near Infrared Camera and Multi-Object Spectrograph (sensitive to infrared light). These give the observatory a good multi-wavelength look at the universe.

HST has made more than 1.4 million observations, looking at planets as close as Mars and Venus to objects out at the limits of the observable universe (about 13.7 billion light-years away). The information from its observations are archived in the Mikulski Archive at the Space Telescope Science Institute in Baltimore, Maryland, USA, and comprises more than 166 terabytes of observational data.

The telescope was built beginning in the late 1970s with funding from NASA and participation from the European Space Agency (ESA). It was designed to do what ground-based telescopes couldn't do: get the highest resolution it could. Ground-based telescopes have typically been limited in resolution (the ability to distinguish individual objects in space) to about 0.5 to 1 arcsecond. If a telescope is orbiting high above Earth's atmosphere, theoretically, it would be possible to achieve higher angular resolution – to get 0.05 arcseconds using a 2.5-metre mirror. So, NASA and the astronomy community forged ahead and began building the new telescope, at first named the Large Space Telescope (LST). It had to be engineered to fit inside the bay of the space shuttle for deployment, and carry a set of science instruments. In 1983, LST was formally named after Edwin P. Hubble, the astronomer who discovered the expansion of the universe from observations of a variable star in the Andromeda Galaxy.

HST Challenges

Once the Hubble Space Telescope was launched and deployed, it began taking images almost immediately. Astronomers, used to working out problems with telescopes, began noticing issues with the sharpness of the telescope's images. After weeks of testing, they came to a very disappointing conclusion: the telescope's main mirror had a condition called 'spherical aberration'. That meant it could not focus light to a sharp point, and thus could not get the high resolution astronomers wanted. It all stemmed from a tiny grinding error, which ended up with the polished mirror being ground wrong by an amount smaller than the width of a human hair.

It was devastating news, but scientists were already hard at work to devise a fix. Solutions included special algorithms to 'deconvolve' the images to take into account the effect of the spherical aberration. As astronomers worked to eke out the science from the telescope's data, planning went on for the first of five servicing missions to the telescope. During the first one, astronauts installed the corrective optics on the telescope and replaced the high-speed photometer. New solar arrays were installed and the mission basically restored the telescope to good working order. The rest of the servicing missions, in February 1997, December 1999, March 2002, and May 2009, regularly repaired aging electronics, swapped out instruments, and – during the last mission – left a fully functioning telescope in orbit. HST should last well into the 2020s.

Hubble's Discoveries

HST gives astronomers deep, high-resolution views into the universe. For example, the famous Hubble Deep Field images were the first to show that the universe contains galaxies as far as we can see, in nearly every direction. These are among the most evocative images ever produced by this orbiting observatory. In astronomy, a deep field is a long-exposure look at a very small region of space; the idea is to gather as much light from objects in that area to get incredibly detailed look at the objects within it. HST is not the only observatory to do deep-field imaging. For example, the ground-based Samuel Oschin Telescope, which was one of the first telescopes to do a large-scale sky survey, was used to take a deep-field image of a section of the Virgo Cluster in 2004 and 2005. The image has thousands of objects in it, and is reproduced on a wall at Griffith Observatory in Los Angeles.

Hubble's Deep Field images probe much further into the universe than any other ground-based images taken at the time. The first one concentrated on a very small area of the sky in the constellation Ursa Major, and the telescope took 342 separate exposures over the course of ten consecutive days in December 1995. The resulting image shows more than 3,000 galaxies, plus some suspected quasars, white dwarfs, and other objects. Astronomers estimated that the view shows galaxies out to a distance of about 12 billion light-years. Hubble has also taken a southern deep field view, as well as what's called an 'Xtreme Deep Field', which extends our view out to a point in cosmic history a few hundred million years after the Big Bang. In 2019, the Space Telescope Science Institute released the Hubble Legacy Deep Field image, combining 7,500 images to show more than 265,000 galaxies. The view 'looks back' to a time about 500 million years after the Big Bang, when the first galaxies were starting to form and evolve.

HST data have also been used to calculate a more precise age of the universe. The telescope has peered into regions surrounding black holes, and directly detected their effects on surrounding material. Astronomers have used it to observe supernova remnants, star clusters, planetary nebulae (the aftermaths of the deaths of stars like our Sun), the births of stars, and most of the planets in our own solar system. HST has also been programmed to search out planets around other stars.

There is enough data flowing from the telescope's instruments to keep another generation of astronomers and their students busy. More than 10,000 papers have been published based on HST results, a wide variety of books have been written, educational videos about its findings are easily found on the Web, and several planetarium shows have been produced. Art gallery exhibitions of Hubble images bring its discoveries to audiences around the world, and some of its iconic images show up on TV shows and in movies. Even fabric makers have gotten 'Hubble vision', printing up specialty material for garments and home decorations using HST images.

Hubble's Siblings: Compton, Chandra and Spitzer

HST is not the only space observatory in orbit these days. In September 2015, it was joined in orbit by the Indian Astrosat, built and launched by the Indian space agency. It covers an astonishing range of light, from optical, infrared, ultraviolet, low- and high-energy x-ray parts of the electromagnetic spectrum. If it detected gamma rays and high-speed particles, it would very nearly be the perfect multi-wavelength observatory!

The age of 'Great Space Observatories' that began with HST also included the construction and deployment of the Chandra X-ray Observatory, the Compton Gamma-ray Observatory (CGRO), and the Spitzer Space Telescope. Of the four, Hubble and Chandra are still functioning at the time of writing. CGRO was decommissioned and deorbited in June 2000, while Spitzer was programmed to transmit its last data in January 2020.

CGRO, which was launched in 1991 and functioned for nearly ten years. It detected photons of light with energies that measure from 30 giga-electron-volts down to 20 kilo-electron-volts. These come from gamma-ray bursts (GRBs) which originate in very distant galaxies. GRBs are the brightest events known in the universe and can happen very quickly (in a period of milliseconds) or last for several hours. The detection of a gamma-ray burst usually sets off a search for an optical counterpart so astronomers can understand what caused it. Interestingly, these bursts were first detected by another set of satellites called Vela. They were built to detect nuclear weapons bursts (usually meaning a nuclear weapons test), but they also uncovered these mysterious brightenings from across the cosmos.

Spitzer Space Telescope began life as the Space Infrared Telescope Facility and was launched in 2003. It was renamed after Lyman

Spitzer, Jr, who was also instrumental in getting Hubble funded and built. The telescope was built in response to astronomers who wanted to be able to study the universe in wavelengths of infrared light that aren't detectable from Earth (due to atmospheric absorption). Its primary mirror is 85 cm (33 inches) across and is made of beryllium. To keep it highly sensitive to infrared light, the unit was cooled to 5.5 K (-267 C). Light comes in to the telescope, hits the mirror and then is fed to three instruments which perform imaging and spectroscopy on light that cannot otherwise be seen on Earth.

Spitzer was launched aboard a Delta II rocket in August 2003. During its years of operation, the observatory spent its time studying star birth regions, hot Jupiter-type exoplanets, the deaths of sun-like stars, dusty galaxies, and other objects and processes that emit prodigious amounts of infrared light.

Spitzer's Exoplanet Search

Spitzer Space Telescope outlasted its original design specifications and is currently in what's called 'Spitzer Beyond'. One of its main aims is the search for exoplanets. It can accurately target stars and is still a very stable spacecraft. Its infrared array camera (IRAC) has been used to find planets using the transit method, which measures minute changes in brightness as a planet transits its star. The dips in brightness can be very tough to detect, and so they need to be measured quite accurately.

Spitzer also used another technique called 'microlensing. It's the phenomenon of gravitational lensing on a small scale and happens when a star passes in front of another star. The gravitational pull of the first star acts as a lens, which magnifies the light from the second star and makes it look brighter. The use of microlensing lets scientists look for a blip in the brightening. If it's there, it might mean there's a planet orbiting the foreground star. This is not a new idea – a ground-based search for such lensing, called the Optical Gravitational Lensing Experiment (OGLE), did the same thing, and Spitzer was combined with the effort.

Spitzer has long been used to look at light from some of the most distant objects in the universe. In particular, the telescope has detected galaxies about 12 billion light-years away. One of the most amazing discoveries it made was in conjunction with the Hubble Space Telescope and a study of a very distant galaxy called GN-z11. Light from the galaxy was first emitted some 13.4 billion years ago – that is from a time only a few hundred million years after the Big Bang. The detection was the first time anyone had seen a galaxy so early in cosmic history, and it tells astronomers that galaxy building processes were at work in the very young universe. With Spitzer's mission at an end, it will next fall to the James Webb Space Telescope to take up purely infrared observations. Data from Spitzer is being used to identify distant targets for Webb when it is finally launched and commissioned, and is now available for astronomers to mine for additional insights into its targets.

Studying the X-Ray Universe with Chandra
The Chandra X-Ray Observatory (CXO) is another of NASA's Great Observatories. It was designed explicitly to study x-rays from high-energy objects of the universe. So, it regularly looks at such events as supernova explosions, peers into the active hearts of galaxies, and studies regions around black holes that are permeated with x-rays. Like other orbiting observatories, Chandra is built to withstand the rigors of space. It has to be precisely aimed during science operations, and faces wild temperature swings and high radiation. It consists of an x-ray telescope, which uses highly polished mirrors to let incoming x-rays ricochet off and toward the detector, and a set of science instruments which record the x-rays.

Planning for the observatory began in 1976, and it was originally called the Advanced X-ray Astrophysics Facility. Its name was later named to Chandra, in honour of the astrophysicist Subrahmanyan Chandrasekhar. His work in determining the ultimate mass of white dwarfs led astronomers to further studies in high-energy phenomena that emit x-rays and other energetic radiation. Chandra was launched in 1999 from the space shuttle Columbia and eventually ended up in an orbit 200 times higher than HST. Chandra is in a highly elliptical orbit which takes it a third of the way to the Moon before returning to approach Earth at a distance of 16,000 kilometres (9,942 miles) every 64 hours. It spends most of its time well above our planet's radiation belts and can observe objects for dozens of hours at a time.

X-ray astronomy covers a wide range of sources that are hot and energetic. They can be as close as the Sun and as far away as the most distant objects in the universe. Neutron star collisions, for example, release prodigious amounts of x-rays. They also unleash gravitational waves, which until very recently, were extremely difficult to detect. Bright bursts of x-rays in the universe can be a clue toward such catastrophic collisions. Neutron star mergers happen relatively quickly, and so the x-ray burst from these objects – often referred to as magnetars – give important clues to direction, distance, and intensity of the merger.

It's not just mergers that give off x-rays. Supernovae are famous for their bursts of highly energetic particles when their massive progenitor stars die. Chandra studies those, too. In addition, material which wanders too close to a black hole can get superheated to extremely high temperatures. Chandra has tracked the remains of a star as it gets sucked into the black hole, watching for the distinct x-ray burst as its material is lost. From its data, astronomers can determine the speed of the black hole's spin. In at least one case, the velocity of the spin was at least half the speed of light.

Some of Chandra's most recent observations have been used (along with observations by the European Space agency's orbiting XMM-Newton x-ray telescope) to probe the distances to more than 1,500 quasars, using ultraviolet light and x-rays streaming from those objects. Astronomers used the distance information to study the expansion rate of the universe at very early times. They discovered from this work that the amount of dark energy (which affects the expansion rate) is growing over time.

Going for Gamma Rays

Gamma rays sound exotic and dangerous, and in many ways, they are. We can't see them, and they can be destructive. They are the most energetic part of the electromagnetic spectrum and have the shortest wavelengths and highest frequencies. These characteristics make them extremely hazardous to life, but they also tell astronomers a *lot* about the objects that emit them in the universe. There are plenty of events and processes which give them off, and they do occur here on Earth when cosmic rays hit our atmosphere and interact with the gas molecules.

Gamma rays are also a by-product of the decay of radioactive elements, particularly in nuclear explosions and in nuclear reactors. However, they aren't always a deadly threat. They can be useful in certain cases; doctors use them to treat cancer (among other things). What astronomers are interested in are cosmic sources of these high-energy photons. For a long time, they remained a mystery to astronomers. They stayed that way until instruments were built which could detect and study these high-energy emissions. Today, astronomers detect these rays from supernova explosions, neutron stars, and black hole interactions. Those events give off incredible amounts of energy, and are sometimes very bright in 'visible' light. Over the past few decades, astronomers have detected extremely strong bursts of gamma rays from various points in the sky. They are a few seconds to a few minutes in duration. However, the bursts seem to occur at very great distances, ranging from millions to billions of light-years away. For those bursts to be seen as bright at this distance must mean that these objects and events must be incredibly energetic in order to be seen from across the universe. They can send out prodigious amounts of energy in just a few seconds—more than the Sun will release throughout its entire existence.

Until very recently, astronomers could only speculate about what caused such massive explosions. However, continued observations have helped to track down the sources of these events. For example, the Swift satellite detected a gamma-ray burst that came from the birth of a black hole more than 12 billion light-years away from Earth. The burst occurred very early in the universe's history.

There are shorter bursts which last less than two seconds, and eventually astronomers linked them to activities called 'kilonovae'. These happen when two neutron stars or a neutron star and a black hole merge. As these incredibly dense, massive objects collide, huge amounts of energy are released, including gamma rays. They can also emit gravitational waves. To 'see' these activities properly, astronomers send specialised instruments to space, so they can detect the gamma rays from high above Earth's protective blanket of air.

Gamma-ray Astronomy

The science of gamma-ray astronomy had its start during the Cold War, when gamma-ray bursts were first detected in the 1960s by the Vela fleet of satellites. Given the paranoia of the times, at first people suspected nuclear explosions here on Earth (or in space). Over the following decades, astronomers looked for the sources of these mysterious pinpoint explosions by searching for optical light (visible light) signals, as well as in ultraviolet and x-ray emissions. The 1991 launch of the Compton Gamma Ray Observatory took the search for cosmic sources of gamma rays to new heights. Its observations showed that GRBs occur throughout the universe and not necessarily inside our own Milky Way Galaxy.

Since that time, the BeppoSAX observatory, launched by the Italian Space Agency, as well as the High Energy Transient Explorer (built and launched by NASA) have been used to detect GRBs. The European Space Agency's International Gamma-Ray Astrophysics Laboratory (INTEGRAL) mission joined the hunt in 2002. It isn't so much a telescope as a collection of detectors tuned to 'see' both high-energy x-rays and gamma rays. Each one is tuned to a specific range. It, with ESA's XMM-Newton spacecraft, is searching out what were once thought to be 'mysterious' bursts in the cosmic distance. It is the first to be able to observe objects as they give off gamma rays, x-rays, and visible light, making it a multi-wavelength space observatory.

Science with INTEGRAL

As with x-rays, gamma rays are emitted from really powerful events and objects. In 2018, for example, INTEGRAL and the XMM-Newton together studied a catastrophic stellar explosion in a distant galaxy. At first astronomers thought it might be simply a powerful supernova explosion. But, as time passed, the explosion grew brighter, rather than dimming as supernovae do. The behaviour of the explosion's source led astronomers to speculate that only a black hole could release such high-energy radiation. What these two observatories could have 'seen' was the birth of a black hole or possibly a neutron star. They are what's left when a supermassive star is destroyed in an explosion. The core compresses down to a ball of neutrons. If the gravity is strong enough, it will collapse to create a black hole. If the object turns out to be either of these possibilities, it will be the first time such a 'birth' process was observed.

INTEGRAL has also observed as a black hole devoured material from its companion star, emitting x-rays and gamma rays along the way. It has also caught emissions from dead stars as they disintegrate, and charted the passage of highly energetic material through a galaxy after a supermassive star explodes.

Fermi and the Energetic Universe

More recently, the Fermi Gamma-ray Telescope has surveyed the sky and charted gamma-ray emitters. It's operated by an international consortium of space agencies and it studies the cosmos in the energy range of 10 keV–300 GeV, which express the energy delivered by the gamma-ray photons as well as energetic particles. Fermi's main interest is in capturing gamma-ray outbursts as early as possible and following the event as it brightens and then dies down. As with its sister satellites which study the energetic universe, Fermi makes detections of supernova explosions, active galactic nuclei, pulsars, and terrestrial gamma-ray flashes, in addition to tracking gamma-ray bursters in the distant universe. As astronomers study more of these outbursts, they'll gain a better understanding of the very energetic activities that cause them. The universe is filled with sources of GRBs, so what they learn will also tell us more about the high-energy cosmos.

Detecting the Ultraviolet Universe

Hot young stars, galaxy evolution, and hot material in nebulae and around black holes comprise much of what ultraviolet astronomy studies. For example, new-born stars give off prodigious amounts of radiation, including ultraviolet (UV). Stellar winds from these stellar babies carve out caverns and gaps in their nebulae. When telescopes study distant galaxies, the distinct UV glow (usually looking blue-ish in optical imaging) gives away the location of star birth regions. At the other end of the age spectrum, very old, hot stars also give off UV, as do the central cores of certain types of active galaxies.

So, which orbiting telescopes are performing ultraviolet observations? Hubble Space Telescope is, as well as the Neil Gehrels Swift Observatory and Astrosat missions mentioned earlier. In addition, several other satellites have UV-sensitive instruments, such as the Solar and Heliospheric Observatory (SOHO) studying the Sun. The Japanese Aerospace Exploration Agency deployed the Hisaki satellite, which is also known as the Spectroscopic Planet Observatory for Recognition of Interaction of Atmosphere (SPRINT-A), in 2013. It uses a spectral instrument to study the extreme ultraviolet light emitted from planets in the solar system, particularly the behaviour of their atmospheres and magnetic fields.

The Chinese Chang'e 3 mission landed on the Moon carrying a 15-cm (5.9-inch) ultraviolet telescope. It has been used to study active galactic nuclei, galaxies, variable stars, binary stars, novae, quasars, blazars and other active regions in space from the surface of the Moon. It is the first long-term 'observatory' on the Moon, and the Chinese space agency has other astronomy missions in mind for the near future. Its follow-up 'twin' is the Chang'e 4 lander and rover which, at the time of writing, is exploring the lunar far side.

The Indian Space Research Organisation launched its Astrosat on 28 September 2015 on a proposed five-year mission to study the universe in multiple wavelengths. It is the first dedicated Indian astronomy mission and part of the country's very ambitious space exploration and astrophysics goals. Astrosat has five instruments sensitive to visible light, near- and far-ultraviolet, soft and hard x-rays. Its main targets for study are x-ray emitters such as binary stars, active galactic nuclei, supernovae, galaxy clusters, and the outer atmospheres of stars.

Testing a Space-based Gravitational Wave Detector Concept
In Chapter 6, we will look at some unusual observatories, including some ground-based facilities which are detecting gravitational waves from objects in space. However, astronomers want to devise ways to study them from space. Can it be done? That's the

question members of the LISA collaboration wanted to answer. LISA Pathfinder (which was a testbed for the planned Laser Interferometer Space Antenna) was launched by the European Space Agency in 2015 to test whether or not such detections could be made from space.

The way the experiment worked was this: once at its Lagrange point orbit (the point near two large bodies in orbit where a smaller object will maintain its position relative to the large orbiting bodies), the LISA Pathfinder placed two test masses in what's called a gravitational freefall while its laser interferometer measured their relative positions and orientation with respect to each other. It was highly accurate and the mission planners determined that such accuracy was good enough to be used on the actual LISA mission. The next step will be to build and deploy the full LISA array in the next decade. It will use the same technique as the Pathfinder mission did, and should be able to capture information about passing gravitational waves.

Satellites and the Sun

Our star is one of the most-studied objects in the solar system and a prime candidate for observations from space. Astronomers have been watching it for years. One of the longest-running solar missions is the Advanced Composition Explorer (ACE) launched by NASA in 1997. It orbits around a Lagrange point, the stable spot which allows it constant access to the Sun perfectly balanced between the tug of Earth's gravity and that of the Sun. The ACE mission has been observing the Sun since then, providing information about space weather and giving warnings about geomagnetic storms.

Space weather is the term scientists use to describe conditions in interplanetary space caused by the solar wind and its effects on Earth's magnetic field, and upper atmospheric layers. Especially strong space weather and geomagnetic storms can have severe effects on spacecraft electronics, the orbits of spacecraft around Earth and other planets, and ground-based communication systems and power grids. ACE and other solar satellites can give early warning about the Sun's activity and how strong it is.

The Solar and Heliospheric Observatory (SOHO), mentioned above, is a mission that has given a long-term view of solar activity.

It was built by a consortium of European industries and research institutions, with input from scientists in the US and Europe. It needs to do constant observations of the Sun from a safe, stable orbit. The main aims of the SOHO mission are to study the Sun's outer layers: the chromosphere, the corona, and the transition regions between them. It does this in several wavelengths and performs remote sensing 24 hours a day. In addition, SOHO monitors the solar wind, and uses specialised instruments to perform helioseismology on the Sun's inner regions. Helioseismology is the study of oscillations of sound waves inside the Sun. By measuring those waves, solar physicists can map its interior. SOHO is equipped with a dozen instruments and specialised telescopes for this work. Interestingly, due to its near-constant study of the Sun, this observatory has also been credited with the discovery of nearly a thousand comets that stray close to the Sun, or sometimes collide with it.

Solar Dynamics Observatory
Our Sun is a dynamic star and solar physicists look to its cycles of activity to understand its characteristics. It changes over long periods of time in several cycles, and those require equally lengthy observing times to gather data about both its long-term and short-term variability cycles.

In 2012, NASA began its 'Living with a Star' programme, which emphasises in-depth studies of our star and how it affects the interplanetary space environment. The first mission in a series for the programme was launched in 2010, called the Solar Dynamics Observatory (SDO). It focuses on the Sun's magnetic energy, how it's converted and released from the Sun, and how it is released as the solar wind and streams of energetic particles. This information, in turn, gives solar physicists a closer look at the variation in the Sun's energy output. The mission also performs helioseismology measurements.

SDO has three scientific experiments which measure the extreme ultraviolet output of the Sun, make images of the chromosphere (the layer of the Sun that gives it yellow colour), and make observations of changes in the solar magnetic field. The observatory is continually pointed at the Sun to accomplish its measurements, and is expected to last until its on-board fuel supplies run out.

The STEREO Mission

As good as SOHO, SDO, and ACE are for solar studies, astronomers get some of their best information about the Sun from stereo imaging. The Solar Terrestrial Relations Observatory (STEREO) mission is a pair of spacecraft that have been in orbit around the Sun since 2006. They provide constant stereoscopic images of the Sun, and since they are essentially in the same orbit as Earth, they can provide images of the Sun that aren't always visible from Earth. Of particular interest are coronal mass ejections, which can send clouds of energised particles and radiation toward Earth in a matter of minutes. Of course, from Earth, we can observe them on the near side of the Sun, but seeing these CMEs on the far side used to require application of helioseismology, which studies oscillations of material in the Sun to get an idea of its interior structure. With STEREO, scientists get views of CMEs from the solar 'far side', which helps them prepare forecasts of space weather to alert power grid companies, airlines, space agencies, and satellite operators.

The STEREO spacecraft each carry five cameras. Two are white-light coronagraphs, one is an extreme ultraviolet imager, and two others are heliospheric imagers that study the space between Earth and the Sun. There are also particles experiments, plasma detectors, and a radio burst tracker.

The Parker Solar Probe

Solar observation missions have all done their observing either from distant orbit or like the Ulysses spacecraft that looped around the Sun from pole to pole from 1994 to 2008. What's needed in solar studies is a spacecraft that can get close to the Sun (at least within a few million kilometres) to get more detail. This is why the Parker Solar Probe was built and launched. It will be the first spacecraft to fly into the low solar corona, a region of the Sun's outer atmosphere. From there, it will gather data about the structure of the plasma which makes up the corona, and chart the magnetic field's role in heating the corona. It will also study the solar wind and uncover the mechanisms sending energised particles out from the Sun.

Parker Solar Probe is carefully protected from the extremes of heat and radiation that the Sun will deliver as it orbits. Its scientific instruments are all tucked away inside the spacecraft behind a special shield. Without such protection, the instruments and communications systems would fry within a few seconds.

The Parker Solar Probe was launched in 2018 and is expected to deliver observational data until 2025. During each of its orbits, it will go past Venus for a gravitational assist to help shrink its orbit. Each time it passes by the Sun, it goes into data gathering mode. After leaving the vicinity of the Sun, the spacecraft then shifts to communicating its data back to Earth.

Rosetta: Europe's Comet Chaser

One of the most exciting satellite chases in recent history took place when the European Space Agency launched the Rosetta mission toward a comet in 2004. It wasn't a direct trek to the comet, however. The spacecraft, which carried cameras and other instruments, as well as a mini-lander, first crossed the asteroid belt before arriving at Comet 67P/Churyumov-Gerasimenko in August 2014. It orbited the comet's nucleus, and eventually deployed a small lander called Philae to settle onto the frozen surface. The lander only lasted a short time and ended up in a shaded area of the comet, but it did manage to send back some data before its batteries died.

Rosetta was the first spacecraft to ever rendezvous with and fly alongside a comet. Others, such as Giotto, had flown past Comet Halley's nucleus in 1986, but none had ever stayed with their comets for long periods of time. It was also the first mission to send a robotic probe to a cometary surface, and it got the first high-resolution images of the nucleus, its surface, and its jets. The Rosetta mission ended when the spacecraft was guided onto the surface of the nucleus. That happened on 29 September 2016 and the spacecraft went silent upon impact.

Although the mission has ended, scientists are still poring over the data returned by Rosetta. Its spectral instrument sent back information about the chemical makeup of the comet, which included carbon-based compounds. The comet's coma (the cloud of gas surrounding the nucleus) had large amounts of oxygen gas. There was a small amount of water ice detected.

During the mission, the cameras recorded almost constant changes on the surface of the nucleus. These increased as the comet got closest to the Sun during perihelion. Surface terrains changed, boulders rolled around, and at one point, a landslide occurred when a cliff on the surface cracked. It was associated with outbursts from the comet spurred by heating during closest approach.

The Rosetta mission enjoyed massive public attention through adroit use of social media by the space agency and the scientists involved. In particular, people were quite interested in the little Philae lander, and it became almost like a mascot for the science teams. Livestreaming of specific events during the mission reached out to millions of people around the world, making the Rosetta mission to Comet 67P/Churyumov-Gerasimenko a truly global phenomenon.

Observing Planets from Space

Ever since the early 1960s, space agencies have been sending spacecraft to other places in the solar system. The Rosetta mission is certainly one good example, but there are many others. Are these observatories? In the strictest sense, yes. They're sent to observe single (or perhaps multiple) solar system targets for flybys or long-period orbital missions, and so that makes them unique in the annals of observations. More than 200 missions from the US, the UK, Europe, Russia, and more recently China, India, and Japan, have been sent to every planet in the solar system, plus dwarf planets, asteroids, and comets. This includes the number of crewed missions that went to the Moon (carrying human explorers). It's an ever-growing number of explorations within just our solar system alone.

The age of *in situ* planetary exploration began at about the same time as the Space Race heated up between the United States and the Soviet Union. The Moon was the obvious first target. There were flyby and lander missions to Mercury and Venus, and then on to Mars. These commenced early the 1960s, and missions to both planets continue today. Some of the latest missions include Akatsuki (Japanese for 'Dawn'), a Japanese spacecraft sent to Venus to study that planet's heavy atmosphere. It was placed into orbit around Venus on 7 December 2015 and in addition to its normal sensing, the spacecraft also detected a massive gravity wave in the planet's atmosphere. Another mission, called Hayabusa 2 (also sent by Japan), arrived at the asteroid 162173 Ryugu, settling in for a multi-year sensing mission.

The second half of the 20th century saw an explosion in planetary exploration by robotic probes. The 21st century continues that work, with in-depth studies of Jupiter, Saturn, the dwarf planets Ceres and Pluto, and the small Kuiper Belt object currently called Ultima Thule. These missions to various solar system bodies have taught us more about our place in space. It's worth taking a brief look at some selected missions to get a general idea of what they've done.

Exploring the Red Planet

Mars has been a tantalising target for study since Schiaparelli and Lowell brought it to public attention at the end of the 19th century. Today, more than seventy-five missions from the United States, the Former Soviet Union, Britain, and India have been mounted to the Red Planet. Many were successful, although many more failed for a variety of technical reasons. There have been flybys, orbiters, mappers, and landers, all of which have returned incredible amounts of data about the planet's surface, atmosphere, and hints about its origin and evolution.

Mars orbiters are the closest thing we have to orbiting space telescopes at the planet, and there are currently several of them studying and mapping the surface and atmosphere. The Mars Express, Mars Orbiter Mission, MAVEN, Mars Reconnaissance Orbiter, ExoMars Trace Gas Orbiter, and Insight are giving continuous long-term information. The Curiosity rover is traversing the surface, returning 'ground truth' data about the mineralogy, weather, and geology of the planet. Other missions are being planned, built, and will be sent in the coming decade.

What is the point of studying Mars so closely? Most people assume it's because humans want to travel there. That's true, but it is not the complete reason. Mars is the closest type of planet to Earth we'll find in the solar system. Understanding how it formed and what influenced it to change from a warm, wet planet to a dry and dusty desert world billions of years ago may supply some clues to how our own planet works. Unlike Earth, however, Mars has lost much of its atmosphere, and orbiters such as the Mars Atmosphere and Volatile Evolution (MAVEN) mission and India's Mars Orbiter Mission (MOM) study the Martian atmosphere for clues to the ongoing loss. What they find may well help planetary scientists figure out the Martian water loss, as well as where the remainder is locked away. If people *do* want to travel there, they need a lot of up-front information about conditions. For example, they need water to survive. The Mars Exploration rovers, Spirit and Opportunity, found very strong evidence there was once was liquid water on the Martian surface. The Mars Reconnaissance Orbiter followed up on Mariner and Viking imaging, which showed tantalising evidence of Mars water in what looked like flood plains and dry riverbeds. Those landscapes hide vast amounts of water in the form of permafrost, and the polar ice caps are known reservoirs of both water and carbon dioxide ice. Future Mars explorers will need all of this is knowledge as they set foot on the planet and make the attempt at humanity's first long-term foothold on another world.

Jupiter Missions

Far beyond Mars, the Juno spacecraft is orbiting Jupiter in a nominal five-year mission to find out more about the giant planet, its atmosphere and magnetic field. It's the second orbital mission to study Jupiter. The first spacecraft to reach the planet were the Pioneer and Voyager missions back in the late 1970s and 1980s. Following that was the long-term Galileo spacecraft which explored Jupiter and returned images and data about the planet. The Juno mission is the latest mission to arrive at and study Jupiter. Each of these missions carried a payload of instruments and imagers to do extensive studies of their targets.

Jupiter is the largest of two gas giants in the solar system, the other being Saturn. It is the second most-massive object in the solar system after the Sun. Dozens of tiny moons circle Jupiter, including the four 'Galilean satellites' first noted by Galileo Galilei through his homemade telescope. The planet also has a tremendously strong magnetic field, a heavy atmosphere, and a very thin set of rings, mostly made of dust particles. Much of what we know about the planet came from spacecraft flybys and the long-term Galileo mission.

Juno's scientific aims at Jupiter are many: it will measure the abundance of water in Jupiter, measure its mass, map the gravitational field, take the planet's temperature, and study the structure of the magnetic field. At the end of its mission, the spacecraft will deliver its final load of images and data, and then enter the Jovian atmosphere, where it will be destroyed. Future missions to Jupiter, and there are some on the drawing boards, will zero in on the small moon Europa. There, specialised instruments will be used to look for traces of life on or beneath the icy surface.

Dawn at Ceres

Solar system exploration doesn't just include trips to the planets and the occasional comet. Planetary scientists have sent spacecraft to other places, such as asteroids and dwarf planets. The Dawn mission, for example, was a planetary observatory sent to orbit the asteroid Vesta for a year, and then continued to the dwarf planet Ceres in the Asteroid Belt. It became the first spacecraft to visit a dwarf planet, just a few months before New Horizons arrived at dwarf planet Pluto.

Dawn mission scientists knew Ceres contained large amounts of ice, likely buried well beneath its surface. It turns out that up to a quarter of this world is made of ice and water. What they didn't expect to find was ammonia deposits, which indicate it probably came from the outer parts of the solar system, where ammonia is more abundant at planets such as Jupiter and Saturn.

Planetary scientists also were surprised to find a towering volcano on Ceres, rising up some 4,000 metres (13,000 feet) above the cratered surface. In the very recent past, a bubble filled with salt water, mud and rock ascended from the core of the planet and flowed out onto the surface. This implies there's a partially fluid, mobile layer of material inside. This is not information that any ground-based observatory could have discovered.

The spacecraft captured visual evidence of ongoing activity, likely just beneath the Ceres surface. The imaging showed Ceres to be heavily cratered from impacts, but also its surface had bright features made of carbonates and other salts. There appears to be a mix of rock and other materials in the surface, as well as patches of ice. It appears this dwarf planet is a layered world (planetary scientists refer to it as 'differentiated'). There is probably a rocky mantle covered by a water-rich crust.

Preparing for Exploring and Mining an Asteroid
One of the future projects that combines planetary exploration with commercialisation is the opportunity, at some point soon, for asteroid mining. Planetary scientists studying asteroids know they could contain resources useful for future exploration. Mining experts look to these objects for commercial ventures. While asteroid mining is probably at least a decade away, sending spacecraft to them has happened in the past and will continue to be a part of solar system exploration. The OSIRIS-Rex mission, sent to space by a consortium of university and resource institutions plus the aerospace industry, is at asteroid 101955 Bennu and is mining this tiny body for samples to return to Earth in 2023.

Bennu is a type of carbonaceous asteroid that used to be part of a larger asteroid or some other proto-planetary body in the early solar system. 'Carbonaceous' means it contains carbon and was formed in the original solar nebula. These asteroids are the most common types. Images from the spacecraft show Bennu has a rough surface with what looks like boulders and large rocks scattered around its surface. The spacecraft's cameras also caught a glimpse of particles scattering off the surface, and dissipating out to space. The studies that OSIRIS-Rex is doing, plus the samples it will send back to Earth for analysis, will give planetary scientists more insight into its formation and evolution since formation of the Sun and planets

Saturn Exploration
The planet Saturn is the epitome of an alien-looking world. Galileo saw it as a round object with 'ears', which the glittering ring system resembled through his crude telescope. Better telescopes showed better views and today, the planet remains one that amateur observers return to time and again for astrophotography. Saturn has long been observed from the ground and recently has been a target for powerful telescopes at Hawaii and other places as astronomers study its atmospheric changes. It has also been a target

of space-based observatories, including Hubble Space Telescope and the Pioneer and Voyager flyby missions.

Cassini-Huygens was the first multi-national spacecraft mission to reach Saturn. It orbited in the system beginning in 2004 and ending in 2017, when the spacecraft was de-orbited into Saturn's upper atmosphere. Cassini carried cameras, specialised instruments to study surfaces and atmospheric chemistry of the Saturnian system, a power source, and communications facilities that relayed data back to Earth. When the spacecraft arrived, it began an in-depth, up-close study of Saturn and its moons, plus the rings. The moons held the most promise of new finds, particularly Titan, the largest moon. The spacecraft dropped the Huygens lander onto Titan, and images and data showed that the moon has a thick smoggy atmosphere and a frozen nitrogen surface dotted with lakes of liquid ethane and methane.

Cassini also sent back tantalising information about Saturn's moon Enceladus. This world is spraying ice particles out from beneath the surface, which indicates the existence of an interior ocean. The mission spent considerable time studying Saturn's clouds and stormy atmosphere, returning images of a fascinating hexagonal-shaped vortex that swirls around over its north pole.

While Saturn isn't the only place with rings, its system is the first and most massive that we've seen. Astronomers suspected they were made mostly of water ice particles and dust, and Cassini instruments confirmed it. The particles range in size from tiny specks of sand and dust to worldlets the size of mountains here on Earth. The rings are divided into ring regions, with the A and B rings the largest. The larger gaps between rings are where moons orbit. The E-ring is made up of ice particles spewing out from Enceladus.

Enceladus is thought to be one possible place where conditions could support life. NASA has referred to it as an ocean moon, and its geyser-like jets indicate there is liquid water under the icy surface. An internal heat source keeps this saltwater ocean in a near liquid and slushy state. That water and heat source could support life. Future missions could help determine whether or not it actually exists deep inside Enceladus.

At the end of its mission, controllers on Earth sent final commands to Cassini to direct it into the Saturn atmosphere. They were concerned that if the spacecraft ran out of fuel, it couldn't manoeuver out of the way of a collision with Titan or Enceladus. If this happened, the spacecraft might contaminate those worlds and possibly have an effect on any life existing there. So, it made a steep dive into Saturn, bringing that productive mission to an end.

Pluto Flyby

Without a doubt, one of the most spectacular planetary missions so far of the 21st century was the New Horizons flyby of dwarf planet Pluto. It made a nine-year journey out to the Kuiper Belt, the region where Pluto orbits, collecting data about interplanetary space as it went. It approached and flew by Pluto in early July 2015, with closest approach on 14 July.

What did New Horizons find? It's important to understand what was expected. Since Pluto orbits so far from the Sun, planetary scientists expected it to be a cold place. Whether it was all ice, or a rocky body with an icy surface was up for discovery. The spacecraft's cameras revealed that icy surface, pockmarked with craters in some places, and paved-over icy plains in others. There are chasms, dark areas covered with a material called 'tholin,' and mountain ranges the height of the Rocky Mountains in the United States. Based on surface analysis, Pluto appears to have some kind of heating mechanism going in deep inside. One scientist described it as a kind of giant 'cosmic lava lamp'.

The mission scientists knew going into flyby that Pluto has an atmosphere which might collapse and freeze out onto the surface as Pluto moves further away from the Sun. As the spacecraft left the Pluto system, it 'looked back' at this world, using the light of the Sun shining through the atmosphere to probe the blanket of nitrogen gas enveloping the planet.

The mission also took an in-depth look at the moons of Pluto, including Charon with its distinctly grey colour and dark pole. The data from the spacecraft will help them understand what the icy components are on its surface, and why it appears to be a frozen world with little of the internal activity Pluto exhibits. The other moons are smaller, oddly shaped, and move in complex orbits with Pluto and Charon.

New Horizons was built to do continued exploration through the Kuiper Belt. It passed by another world, called Ultima Thule (2014 MU69) at the end of 2018, and may yet encounter another Kuiper Belt Object, if one can be found along its path.

Exploring Earth from Space

One of the most profound subjects of study in the solar system is our own planet Earth. Scientists have been exploring it since the first Sputnik satellite went into orbit in 1957. Remote sensing of our own world for scientific purposes covers a range of investigations using both imaging and spectral instruments, radio experiments, and other detectors. Orbiting sensors and telescopes monitor our climate, meteorology, overall environment, track seasonal changes on the continents and oceans, and in more recent times, help scientists monitor such things as changing sea ice packs and the rise of CO_2 levels in our atmosphere. Nearly 700 such observing platforms orbit the planet. That doesn't include spy satellites and commercial communication platforms. When those are added, the number is close to 2,000 and rising.

With the addition of a new generation of small-sized CubeSats, which are already in use in growing numbers, the collections of Earth-observing platforms will continue to grow. CubeSats are typically very small cube-shaped spacecraft which carry very simple instruments to space. They are launched by a variety of users, including private industry, college students and even some high school students in cooperation with NASA educational programmes.

NASA's Earth Observing System satellites provide a good look at the kinds of work being done by science-oriented observatories that have Earth as their main target. The EOS began with the launch of the ATS-3 weather observation satellite in 1966. Since then, the agency has sent spacecraft to study interactions between Earth's magnetic field and the solar wind, to monitor Earth's radiation budget (that is, how much energy it absorbs and releases), to study atmospheric chemistry, measure ocean waves and movements, to assess how humans are impacting the environment, to map the global surfaces, to collect information about the planet's water cycle, to chart changes in cloud cover during the seasons, to measure plant growth on the continents, and many scientific studies. Some components of the Earth Observation System also flew aboard shuttle missions. All of these studies have been correlated with ground-based studies, as well.

There's no doubt space-based observing platforms – whether they are orbiting at or near Earth, looking out from Earth or back *to* Earth, studying the larger universe, or visiting other worlds of the solar system – provide images and data which help planetary scientists understand the worlds orbiting the Sun and the stars and galaxies beyond our own Milky Way. They also give astronomers who study exoplanets a template for how planetary formation happened around our star, and how it results in worlds around distant stars. All the space-based observations have literally changed our views of the universe and will continue to do so into the future.

6

UNUSUAL OBSERVATORIES AND CELESTIAL TARGETS

Traditional astronomy looks at the stars, planets, galaxies, and clusters of galaxies that make up the universe. As we've seen in previous chapters, these are well covered by both ground-based and space-based telescopes. In this chapter, our tour moves on to some of the more unusual observatories that exist on the planet, move high up into the atmosphere, or orbit in space. And, we will look at a few which didn't make it past the drawing boards, failed in some other way, and in at least one case, examine one with a reputation it didn't deserve.

These unusual facilities exist because there are other objects and events in the universe which require specialised detectors that aren't quite like the observatories we've learned about so far. What kind of targets? Our Sun is one, and there is a long tradition of solar telescopes both on Earth and in space. Observations in various wavelengths of light such as infrared or x-ray were also better done high in the atmosphere (or, better yet, in space). Then, there are the neutrinos, cosmic rays, and gravitational waves. They all require specialised observational tools and techniques.

High Flight, High Astronomy
In Chapter 5, balloon-based observatories were mentioned as predecessors to space-based astronomy missions. Scientists wanted

to get their instruments well above the atmosphere into regions where they could make infrared and x-ray observations. Such launches continue today, many from Antarctica, deployed by NASA's Balloon programme, and projects by other institutions around the world. In 2018, for example, Washington University in St Louis, Missouri, launched its X-Calibur instrument aboard a helium balloon. Its flight took it above 99 per cent of Earth's atmosphere and the mission was aimed at measuring polarisation of x-rays from neutron stars, black holes, and other active events in the universe. Some balloons carry instruments that measure conditions in Earth's atmosphere, as well.

The South Pole isn't the only launching pad for balloon-based astronomy. In late 2018, Arizona's World View Enterprises sent its Stratolite high-altitude balloon system up from a remote area in McCall, Idaho. It carried a specialised payload called the High-Altitude Electromagnetic Sounding of Earth and Planetary Interiors instrument. The goal of the mission was to measure how electromagnetic waves can penetrate the surface of a planet and what its information might reveal about the target. Such an experiment could be applied at other worlds with atmospheres, particularly Venus and its thick blanket of gases.

One notable experiment that 'starred' in an award-winning film was the Balloon-borne Large Aperture Submillimetre Telescope (BLAST). The instrument itself is a 2.5-metre (8.2-foot) telescope which focuses incoming light onto an array of cryogenically cooled detectors. The whole assembly is carried up into the stratosphere by a balloon and gondola. The instrument has made three flights. The first was an engineering test from a site in New Mexico in 2003. Another was launched from Kiruna, Sweden, in 2005. The third mission (and the second science-gathering one) was deployed from McMurdo Station in Antarctica in January 2007. The fourth science mission lifted off from McMurdo in 2010.

All three collected data successfully, but the third deployment ended in disaster. Upon landing, the telescope was dragged along on the ice for a full day and got wedged in a crevasse about 200 kilometres (124 miles) away from the originally targeted touchdown spot. The drives were retrieved and the data recovered.

That recovery was highlighted in a video called *Blast!*, made by Paul Devlin, whose brother is one of the scientists on the mission. More missions are scheduled in the near future.

BLAST scientists use the telescope to study magnetic fields both in the Milky Way galaxy as well as in nearby galaxies. The instrument is also studying and mapping distant star-forming clouds to see how magnetic fields affect the birth and evolution of young stars. In addition, it studies dust in the space between stars (the interstellar medium) to see how interstellar magnetic fields influence the density and distribution of such clouds.

Airborne Telescopes

It wasn't long before astronomers began looking toward high-flying aircraft as possible platforms for telescopes. Although there were some space-borne telescopes (such as the Orbiting Solar Observatories), getting to space was and remains an expensive and risky proposition. Aircraft offered a less-costly approach. Of course, spy planes had flown at high altitudes for years, but they were photographing objects on Earth's surface, not looking out to space. The first attempt at a flying observatory was a test of a specially outfitted Concorde plane. It was done to allow astronomers to photograph and study the solar eclipse of 30 June 1973. It was proof that such an observatory could be built, provided it could be made stable and reliable.

In 1974, the first successful purpose-built flying observatory took to the skies. It was the Gerard P. Kuiper Airborne Observatory, named for a prominent astronomer known to many as the father of planetary science. It carried a 91-cm (35-inch) reflector optimised for infrared observations. The aircraft was a modified C-141A jet and it could travel up to 14 kilometres (9 miles) altitude and allowed astronomers to perform infrared observations well above the water-rich atmosphere. KAO, as it was called, flew 1,417 missions and was retired in 1995. During that time, it enabled the discovery of the rings of Uranus in 1977, the atmosphere of Pluto in 1988. Scientists used it to look at disks around nearby stars, and made observations of water and organic molecules in star birth regions. It also studied the centre of the Milky Way galaxy, where a supermassive black hole is now known to exist. In 1987, shortly after the appearance of supernova 1987a in the Large Magellanic Cloud, KAO's telescope studied it and provided information about the formation of heavy elements in the cloud of debris rushing away from the exploded star.

After KAO was retired, work began on another airborne observatory called the Stratospheric Observatory for Infrared Astronomy (SOFIA), also operated by NASA in cooperation with the German Aerospace Centre, DLR. This observatory is based on a Boeing 747 aircraft and carries a 2.7-metre (8-foot) reflecting telescope. SOFIA's first-light observations took place on 26 May 2010, and it is still flying. Its primary mission is to study planetary atmospheres, comets, the interstellar medium, and the process of star formation. It has been used to detect oxygen in the thin Martian atmosphere, and it has mapped dust clouds in the region near Sagittarius A*, the black hole at the core of the Milky Way.

On 27 July 2018, SOFIA was used as part of an effort to observe a stellar occultation in the Kuiper Belt. An occultation occurs when an object passes 'in front' of a more distant star. Continuous observations

The Stratospheric Observatory for Infrared Astronomy (SOFIA) with its sliding door open to start observations with the 17-ton (15,422-kg) infrared telescope. Courtesy NASA/Jim Ross.

of the passage help astronomers determine the object's shape. In this case, the object was 2014 MU69. This little world, nicknamed Ultima Thule, was the second flyby target for the New Horizons spacecraft, which encountered Pluto three years earlier, in July 2015. Data from that observation and a collection of ground-based measurements were used to target the spacecraft more precisely for its late 2018 flyby. The ground-based measurements were quite unique. They were made to help New Horizons team members get a good idea of the size and shape of Ultima Thule. In addition, it was important to know if this tiny object had rings or dust clouds nearby. If so, these could pose a threat to the spacecraft as it flew by.

So, the team assembled groups of observers and equipped them with telescopes and sent them to places on Earth where the line-of-sight between them and their target would be in line with the background star that the object would occult. Hubble Space Telescope was also programmed to look at this tiny Kuiper Belt object at the same time. The first set of observations, made in 2017, weren't successful. But, in 2018, two teams of observers went to Senegal and Columbia and were able to capture the occultation and get more information about the size and shape of the target. The accompanying HST observations made at the same time were able to study the region very close to Ultima Thule and show that there were no dust clouds or rings.

Solar Helioseismology: Studying Inside the Sun from Earth

In previous chapters, we learned about ground- and space-based solar observatories. There is another astrophysics discipline which seeks to understand the Sun's internal structure. This is something that really can't be easily determined from some other types of solar observations. So, solar physicists take 'seismic' measurements of the Sun in a technique called helioseismology.

Solar physicists have a good general idea of how the Sun's interior is structured. The Sun is a ball of plasma with a core at its heart. The core is an extreme place. Temperatures there reach 15 million degrees Celsius, and the pressure is equivalent to 250 atmospheres. Those extremes of temperature and pressure are instrumental in fusing hydrogen into helium. The process gives off radiation in the form of heat and light. Both are carried out to the surface of the Sun through a radiative zone, a convective layer, the photosphere, the chromosphere and the corona.

It's been known for some time that the Sun 'rings', or resonates, almost like a bell, with millions of distinct sound waves. Those waves affect the light shining from the Sun's surface; that effect is called 'Doppler shifting'. It means the light's wavelengths are shifted by the unseen waves. Each of the waves moves at different speeds and through different depths of the Sun's layers. Helioseismology detectors let solar physicists probe the temperature, motions of material, and the chemical composition of the Sun from a point just beneath the surface down to the Sun's core (where hydrogen fusion takes place).

Studying the Sun's interior also gives physicists a look inside other stars, by default. Helioseismology allows them to measure the interior structure of the Sun as well as its motions. Scientists have a good theoretical idea of what it's like inside the Sun and can apply what they know to asteroseismology – the study of the interiors of other stars, particularly those similar to our own. They know that hydrogen fusion at the core results in the creation of helium atoms, and a release of energy. All of the energy eventually makes its way to the solar surface through convection currents. Yet, those can't be directly measured. Helioseismology can help measure them and the way they oscillate. It can also be used to study the origins and evolution of sunspots. These are relatively cool areas of magnetic activity that ride the visible surface of the Sun.

NSO and GONG

How do solar physicists perform helioseismology? One way is through the Global Oscillation Network Group (GONG), managed by the US National Solar Observatory. This is a network of six specialised observatories located at Big Bear Observatory in California, the Cerro Tololo Inter-American Observatory in Chile, Learmonth Solar Observatory in Australia, Mauna Loa Observatory on the Big Island of Hawaii, Observatorio del Teide in the Canary Islands, and Udaipur Solar Observatory in India. Each of these network lobes operates stable velocity imagers to obtain nearly continuous observations of the Sun's '5-minute' oscillations, or pulsations.

The Big Bear site, which is a member of the network, is located on a small island in Big Bear Lake, not far from Los Angeles, California. That isolation plus good weather conditions at the site allow a clear, steady view of the Sun nearly year-round. Big Bear's GONG site specialises in very high-resolution solar observations in both visible and near-infrared wavelengths. Each GONG installation has a pair of mirrors which tracks the Sun and feed sunlight into a cargo container. Instruments and filters inside the box allow for different observations. There is also a CCD camera that is used for imaging during observations. The entire set-up is computer-managed and operated, and data is collected every clear day.

Another helioseismology network, called the Birmingham Solar Observations Network (BISON). This network has been in operation since 1975 and has added new sites throughout its history. Data from its network locations at Mount Wilson in California, Las Campanas Observatory in Chile, Observatorio del Teide in the Canary Islands, the South African Astronomical Observatory in Sutherland, the OTC Earth Station Carnarvon, in Western Australia, and the Paul Wild Observatory in Narrabri, New South Wales, spans three solar cycles.

Helioseismology is also done from space, and has been part of the Solar and Heliospheric Observatory (SOHO) long-term mission as well as the Solar Dynamics Mission. On SDO, the main instrument for such studies is called the Helioseismic and Magnetic Imager (HMI). It measures motions of the solar photosphere, the visible surface of the Sun. From those studies, solar physicists can detect the solar oscillations and make measurements of the polarisation of light in specific spectral lines. This gives them insight into the main components of the magnetic field in the photosphere. From that, they can deduce information about the state of the Sun's interior, and also the Sun's variability. Such variability is what causes space weather, which affects Earth and the other planets. The data from the HMI help establish a link between what's happening inside the Sun and its varied magnetic activity. In the long run, it also helps improve information about solar storms and other outbursts, which can affect humans and technology on Earth.

The Sun has fascinated observers throughout history, and with the advent of spectroscopy in the 19th century and subsequent interest in sunspots, the solar cycle, and solar variability, it has remained an important area of study. Today's instruments, space-based and ground-based, are continuing to feed that fascination and also contribute prodigiously to helping us all learn what it means to live with a star.

Neutrino Science and Energetic Particles

In school, most of us learn about atoms, the basic particles that make up the universe. Atoms are themselves made up of smaller bits called 'subatomic particles'. The basic atomic components are protons (which carry a positive charge), neutrons (which have no charge), and electrons (which carry a negative charge). The protons and neutrons are the centre of the atom, while the electrons orbit the nucleus. Essentially, these make up all the normal matter in the universe: you, me, the car you drive, the food you eat, the rain that falls on our heads, the star we orbit, and the other planets, stars, and galaxies.

In the sub-atomic world, there are also neutrinos. These are similar to electrons, but have incredibly small mass and no electrical charge. Neutrinos are everywhere in the universe and are the most abundant particles around. There are three types of these odd particles: the electron neutrinos, muon neutrinos, and tau neutrinos. It turns out they can oscillate between these different types as they travel.

Where do neutrinos come from in the universe? There are a number of sources from across the universe. Astronomers know they are produced in supernova explosions. When a supermassive star explodes, it sends these tiny particles rushing out incredibly fast, at very close to the speed of light. Often, a rush of neutrinos alerts astronomers to the fact that a supernova has exploded. They reach Earth before the light from the catastrophic event can get here.

The Sun also produces neutrinos, although not as many as expected. The shortage of solar neutrinos has puzzled solar physicists for a long time. They even have a name for it: the solar neutrino problem. The question is: why does it seem to produce so few? One possible reason is that the Sun *is* creating neutron neutrinos but they change into muon or tau neutrinos as they travel through space. This might be one explanation for the puzzling lack of solar neutrinos detected here on Earth.

Neutrinos are also produced inside our own planet, from the decay of radioactive materials. They can also be created right in our own atmosphere. This happens when a cosmic ray (more about those below) crashes into the atmosphere. The cosmic rays that don't pass right through smash into atoms in Earth's upper atmosphere. That collision produces what's called a cascade or 'air shower' of secondary particles called protons, neutrons, pions, muons, photons, electrons, and positrons. Pions can then decay further to produce neutrinos and antineutrinos. Neutrinos were also generated by the Big Bang, and some scientists postulate they could be part of the dark matter portion of the universe. In general, however, the neutrinos that intrigue astronomers are those from the Sun and the distant universe.

Research into Neutrinos

Of course, not all neutrinos come from celestial sources. On Earth, neutrinos are emitted by nuclear reactors and are created in beams in particle accelerators, such as the facilities at CERN in Switzerland. Physicists study these 'captive' or 'local' neutrinos to understand and predict their behaviour, while other research is aimed at measuring their properties and masses. Physicists take advantage of the known sources on Earth to put neutrino detectors near nuclear plants, since they emit a constant supply. Reactors create neutrinos through the process of fission, while accelerators collide protons with a target, which ultimately results in a beam of neutrinos. It turns out that neutrinos, while being very tiny and almost massless, are numerous enough they can have an effect on other matter in the universe. This has led some researchers to consider them as dark matter candidates. Future neutrino experiments, such as CERN's Long Baseline Neutrino Facility, will give further insight into the characteristics of this particle.

Cosmic Rays

There is a second type of particle that astronomers look for, known as the cosmic ray. It's something of a misnomer, since it's really a particle. Cosmic rays come from various sources, mostly from outside our planet's atmosphere. The Sun generates some, but most come from supernova explosions, and some fraction of the cosmic rays may also come from active galactic nuclei. These are energetic regions at the hearts of galaxies and are likely powered by black holes. Until recent decades, it wasn't completely clear what other processes and events in the universe would create cosmic rays. This is why scientists set out to study them and learn about what propels. Cosmic rays travel across space and can pass though most matter. They can also damage electronics, and may well induce mutations in living things. These fast-moving particles were first detected early in the 20th century, and have been studied ever since. There are three basic types of cosmic rays: galactic (that is, from sources inside our galaxy, extragalactic (from sources outside the Milky Way), and solar energetic particles, which are mostly protons emitted by the Sun. When cosmic rays smack into Earth's atmosphere, they get converted to secondary particles.

The field of neutrino astronomy is very active, and astronomers have confirmed neutrinos from the Sun, Supernova 1987a (which lies about 160,000 light-years away from us in a companion galaxy called the Large Magellanic Cloud), and a blazar called TXS 0506+056. It's a massive black hole at the heart of galaxy that lies about 3.7 billion light years away. Superheated material is funnelling away from the black hole via a jet.

Modern Methods of Detecting Neutrinos and Cosmic Rays

Spotting and measuring these fast-moving, nearly mass-less particles isn't an easy task. They don't interact very easily with regular matter, which makes them difficult to pin down. Neutrinos can travel through many light-years of space before interacting with interstellar gas and dust, or a planet, or star. They also pass almost completely unimpeded through normal matter (such as a planet or your body). Their gravitational force is nearly impossible to measure, so scientists have to devise methods to find the weak interactions they do have.

For these reasons, neutrino detectors have to have a large 'collecting area' to detect enough for study. These observatories are usually built underground, to isolate the detector from local sources of radiation as well as cosmic rays. Detection requires extremely sensitive equipment and even the best ones on Earth only measure a relative few.

There are several different types of neutrino observatories. Some use a fluid called tetrachloroethylene, which (to non-astronomers) is the main ingredient in dry cleaning fluid. When a neutrino hits a chlorine-37 atom in the tank, it converts it to an argon-37 atom. That interaction can be detected by instruments in the observatory. Scientist Ray Davis created a test lab neutrino detector in one level of the Homestake Mine in South Dakota (in the US, discussed below) and used it to detect solar neutrinos in the 1960s.

Another way to measure neutrinos is through what's called a Cherenkov detector. The name refers to Cherenkov radiation, which is emitted whenever charged particles such as electrons or muons are moving rapidly through water, heavy water, or ice. The charged particle generates this radiation as it transits the detector fluid.

The first neutrino detector experiments from astrophysical sources took place at Homestake. It was actually a working gold mine until 2002, reportedly the largest and deepest of its kind in North America. After it closed, the owners were ready to shut it down and move on, but scientists saw a different kind of use for it: as a full-time neutrino detector and science lab. The mine had all the attributes they needed for such a project: the interior was surrounded by rock, which would filter out cosmic rays. It was self-contained, and they could re-outfit it to run as a science lab. After protracted negotiations with the National Science Foundation, and a temporary shutdown of the mine, it became home to what is now called the Sanford Underground Research Facility and it has been home to several major experiments over the years. One was the Large Underground Xenon (LUX) experiment. Although it is now decommissioned, LUX turned out to be a very sensitive dark matter detector. Another project is called CASPAR, which stands for Compact Accelerator System for Performing Astrophysical Research. It is a low-energy particle accelerator helping scientists understand how elements are created in the universe, specifically inside stars, and how much energy such a process gives off.

The Deep Underground Neutrino Experiment (or DUNE) follows up on the earlier work of Dr Davis. It will use two detectors to record particle interactions at the source of a neutrino beam (created at the Fermi National Accelerator Laboratory in Illinois) and another, larger detector installed deep inside the Sanford lab called the Long-Baseline Neutrino Facility. Scientists hope to make discoveries that uncover the origin of matter, help explain the formation of black holes, and search for a process called proton decay.

A Super-detector in Japan

The Super-Kamiokande neutrino observatory is located under a mountain near Hida, Japan. Like the Homestead detector, this one is also buried deep within a mine, some 1,000 metres below ground. Its stated mission is to detect high-energy neutrinos from the Sun, atmospheric decay, and energetic events such as supernova explosions. Super-Kamiokande uses a large tank filled with 50,000 tons of ultrapure water to 'capture' data using 13,000 photomultiplier tubes (which convert the photons generated into data). When a neutrino smashes into the detector, it creates Cherenkov radiation. That 'signature' is captured and the data analysed to understand the source and energy of the neutrinos.

More Neutrino Labs

In Canada, the Sudbury Neutrino Observatory (now called SNOLAB) operates in a similar way to Super-Kamiokande, but instead of ultrapure water, it uses heavy water (water with deuterium molecules rather than simple hydrogen) to detect neutrinos. SNOLAB's detector is 2 kilometres (1.2 miles) deep, and until the Jinping Underground Laboratory was built in China's Sichuan province, was the deepest in the world. At Jinping, the detector is 2.4 kilometres (1.5 miles) below ground, and is host to a variety of experiments, including the China Dark Matter Experiment, PandaX (another dark matter experiment).

IceCube Finds a Blazar

In the hunt for neutrino sources, physicists and astronomers turn their attention to wherever and whenever a high-energy event occurs. In mid-2018, scientists using the IceCube South Pole Neutrino Observatory announced they had found tantalizing evidence of a very powerful source of high-energy cosmic neutrinos from a source some four billion light-years from Earth. The energetic object is what's called a 'blazar', an elliptical galaxy with a rapidly spinning black hole at its heart. In this case, the blazar is called TXS 0505+056, and it sent out a burst of radiation that resulted in a shower of neutrinos.

When the telescope detected a burst of neutrinos on 22 September 2017, it triggered a worldwide alert containing sky coordinates of the object where the neutrinos seemed to emanate. Other observatories, including NASA's orbiting Fermi Gamma-ray Space Telescope and the Major Atmospheric Gamma Imaging Cherenkov Telescope, or MAGIC, in the Canary Islands, were immediately put into service. They detected a flare of high-energy gamma rays associated with TXS 0506+056, thanks to the 'multi-messenger' approach to observing events with multiple telescopes and observatories.

IceCube Neutrino Observatory looks like a futuristic lab, set in a science-fictional landscape on an alien planet. In reality, it's situated at the South Pole and its detector is actually a cubic kilometre of ice, threaded with 86 cables that each carry 60 digital optical modules. They are frozen into the ice, which is optically clear. The observatory is administered and used by an international group of scientists. It's funded by the United States National Science Foundation, as well as agencies in Asia, Canada, England, and Europe.

The Antarctic neutrino observatory, which also includes the surface array IceTop and the dense infill array DeepCore, was designed as a multipurpose experiment. IceCube collaborators address several big questions in physics, such as the nature of dark matter and the properties of the neutrino itself. IceCube also observes cosmic rays that interact with the Earth's atmosphere, which have revealed fascinating structures that are not presently understood.

The Pierre Auger Array Cosmic Ray Detector

While neutrino observatories don't necessarily want to 'see' cosmic rays, there are places which study these energetic, massless particles as well. The Pierre Auger Observatory was named after the physicist who discovered the showers of particles that result when cosmic rays hit our atmosphere. A site in which a series of detectors were linked together was built near Malargüe, Argentina. It is used by physicists from nearly a hundred research institutions around the world and comprises a series of detectors dotted around a detection area measuring 3,000 square kilometres. This unique layout maximizes the numbers of cosmic rays that can be collected.

The Auger Observatory is uses two independent ways to detect and study them. One technique detects high energy particles through their interaction with water placed in surface detector tanks. Each tank holds 12 tons of ultrapure water. When particles in a cosmic ray shower race through the tank the subsequent collisions emit light as Cherenkov radiation. Those photons are picked up by photomultiplier tubes. The Auger Observatory's detectors are incredibly sensitive, many times more perceptive than our eyes. They can watch as distant air showers develop in the upper atmosphere.

A Space-Based Cosmic Ray Detector

Up until 2011, long-term particle physics experiments took place on Earth. Of course, orbiting spacecraft have recorded the passing of cosmic rays, but none were specifically built for the purpose. This changed with the launch of the Alpha Magnetic Spectrometer-2, a particle physics detector currently installed on-board the International Space Station. It is making measurements of cosmic rays and the ultimate aim is to provide insight into the nature of dark matter, antimatter, and missing matter. During its first eight years in orbit, the detector has collected nearly 140 billion cosmic rays. It is also monitoring the near-Earth space environment for cosmic rays, which can and will pose a threat to long-term human habitation of space. In particular, future researchers living on the Moon, or in transit to Mars or an asteroid, will face increased risks, not just from cosmic rays but from other types of radiation, as well.

Cosmic Ray Truth from VERITAS

The most energetic cosmic rays have caused scientists to wonder just what would be so powerful to accelerate these particles to velocities near the speed of light. As mentioned above, supernova explosions have long been implicated in launching cosmic rays across the gulfs of space. The very rarest high-energy cosmic rays carry more than 100 billion times the energy as is generated by any particle accelerator on Earth.

The Very Energetic Radiation Imaging Telescope Array System (VERITAS) system on Mount Hopkins in Arizona is a set of four 12-metre (39-foot) telescopes that detect very high-energy gamma-ray photons from space. VERITAS is one of four such arrays; the others are the High Energy Stereoscopic System (HESS), in Namibia; the Major Atmospheric Gamma Imaging Telescopes (MAGIC), in the Canary Islands; and the First G-APD Cherenkov Telescope (FACT), in the Canary Islands. This collection of detectors, looking a lot like radio telescopes, measure flashes of Cherenkov radiation given off when a high-energy gamma ray hits Earth's atmosphere and produces a cascade of charged particles. Since they are spread out over a greater area than space-based gamma-ray detectors can cover, such instruments can supply a greater number of detections.

In 2009, VERITAS found evidence for very high-energy particles which exist in a distant galaxy called Messier 82, the so-called 'Cigar Galaxy'. It lies at a distance of twelve million light years and appears to be undergoing a great deal of activity. M82 is known as a prodigious star-formation galaxy, undergoing substantial starburst activity. Like other starburst galaxies, it also has a large number of massive stars and supernova explosions. Such explosions would create swarms of cosmic rays. VERITAS confirmed that M82 has more than 500 times the average number of cosmic rays than our own Milky Way, which has much less star birth activity. However, most of those cosmic rays remain trapped inside that galaxy. So, how to detect their presence? The VERITAS instrument didn't directly look for them. Instead, since it's a gamma-ray detector, it looked for those. When cosmic rays in M82 interact with interstellar gas and radiation, they produce gamma rays, which VERITAS and other Earth-based gamma-ray detectors seek out. The amount of gamma rays, and the energies measured, reveal more information about their source, even across millions of light-years of space.

The Hunt for Dark Matter With the XENON Project

Astronomers have long sought an explanation for dark matter particles and a new generation of particle detectors may provide answers. A number of candidates for dark matter have been suggested, including neutrinos. But, for many years, a particle called a WIMP -- which stands for "weakly interacting massive particle" -- has been suspected to be a component.

The XENON project, which is operated in Italy at the Gran Sasso Laboratory, is attempting to detect these elusive WIMPS. It uses a chamber filled with liquid xenon buried deep underground. The project looks for scintillation and ionization reactions when dark matter particles interact with each other. XENON is a collaborative effort between institutions across the world and is currently being updated to facilitate enhanced data-gathering. Along with telescopes and space-based detectors, projects such as XENON will pave the way to an understanding of what dark matter actually is.

On the Hunt for Gravitational Ripples

Gravitational waves have been much in the news over the past several years as the first successful detections of these disturbances were announced by the consortium of scientists running the Laser Interferometer Gravitational-wave Observatory (LIGO) in the United States. While there is a space-based observatory for these events, all the current discoveries have taken place on Earth.

What are gravitational waves? Simply put, they are ripples in the fabric of space-time. Most readers are familiar with that term from Albert Einstein's work in special relativity. The result of his theories is a model that brings together the familiar three-dimensional universe we know with the the dimension of time. This 'fabric' of the universe can be affected by gravity and perturbations introduced by gravitational forces. This is where gravitational waves come in. According to Einstein's mathematics, very massive objects which are accelerating can disrupt space-time. As they do, they generate waves of distorted space that would radiate away from the object, or objects. As the waves move away, they travel at the speed of light, and they are carrying along information about their sources.

Very strong gravitational waves are created when very catastrophic events occur. This almost always includes the actions and interactions of very massive objects. Colliding black holes radiate gravitational waves, as do merging neutron stars, and the collapse of massive stars as supernovae. There is even some evidence there were gravitational waves created in the Big Bang, some 13.7 billion years ago.

So, how do scientists go about detecting these ripples? It turns out there have been 'indirect' detections of them since 1974, when astronomers found a binary pulsar that had two stars orbiting each other and getting closer over time. According to Einstein's predictions, this type of system should have been sending out gravitational waves, although they hadn't been detected directly. Instead, astronomers noted the stars were closing in on each other at a rate consistent with a system giving off gravitational waves. That's because the waves would remove energy from the orbiting pair and cause them to move closer together. Other binary pairs have shown the same decay in their orbits, and this led astronomers to deduce that gravitational waves were at work.

Building a Gravitational Wave Observatory
Creating a laboratory to detect gravitational waves, especially here on Earth, is a painstaking process. Since it measures minute perturbations due to gravitational waves, the detectors have to be incredibly sensitive. Scientists have to work to filter out sources of interference that could be mistaken for a gravitational wave (or mask the passage of one). Something as simple as a heavy vehicle driving on a nearby motorway can affect readings.

Scientists set out to build LIGO in the late 1980s, after years of experiments. It was finished and began operating in 2002. It didn't detect any waves, and was shut down and enhanced with new hardware and software. LIGO reopened in 2015 and began taking measurements again. LIGO itself is a laser inteferometer in two locations: one in Hanford, Washington, and the other Livingstone, Louisiana, in the US. Each facility has two 4-km-long (2.4-mile) 'arms' in the shape of an 'L'. Think of them as 'antennae' whose only purpose is to detect gravitational waves, and are at once both an observatory and a state-of-the-art physics experiment.

How a Gravitational Wave Interferometer Works
The LIGO design is based on what's called a Michelson inteferometer. It was invented by physicist Albert Abraham Michelson in the 1880s to measure patterns and intensity of a light beam that has been split and recombined. The result is an interference pattern which can be measured and analysed. Michel interferometers are often used to make very small measurements that can't be done using other methods. It also makes them very good at detecting gravitational waves.

In the LIGO interferometer, a laser beam moves through a 'beam splitter,' which divides the original single beam into two separate beams. Half the light passes through, while the other half is sent off at a 90-degree angle. Each beam travels down an arm of the interferometer, through vacuum chambers. When the beams reach the ends of the L-shaped arms, they bounce off mirrors and head back on their initial path. They meet at the beam splitter and are merged back into a single beam. When they meet, their waves interfere with each other and then both are sent to a photodetector. If the arms don't change length while the beams are making their trip, then the light waves in the recombined beam cancle each other out. When that happens, no light actually reaches the photodetector.

However, if the arms *do* change length, that is, if the distance traveled by the split laser beams does change, then one beam takes longer to make the trip than the other. This tiny change of length – and it *is* tiny, only 1/10,000 of the width of a proton – is made when a gravitational wave passes by. Both detectors have to 'see' this shift for it to be considered a viable gravitational wave candidate. When such a wave passes by, and the light paths are merged, the recombined light waves are offset by a tiny amount, which produces interferences. At that point, the detector sees some light. This is a highly simplified look at how LIGO 'sees' gravitational waves. Essentially, its detectors look for very specific characteristics (how the interference pattern changes over time).

This is basically what happened on 14 September 2015. That's when LIGO sensed a gravitational wave. It turned out to be distortions in spacetime generated by two colliding black holes. They lay nearly 1.3 billion light years away. The detection manifested itself as a 'chirp' in the signal seen by both the Hanford and Livingston LIGO instruments. The signal lasted 0.2 seconds and changed in frequency and amplitude as the signal continued. Those changes were the signals emitted as each object orbited faster in the final seconds before the merger took place. Not only did the chirp tell scientists a merger caused the waves, but they could even figure out the orbital speed of both objects was about 60 per cent of the speed of light at the moment they collided. So, not only did the gravitational wave detection tell something about the objects themselves, but it revealed the tremendous energy of the collision. Such information provides a very new and exciting way to probe distant objects and events in the universe.

VIRGO and KAGRA

LIGO is not the only gravitational wave detector on the planet. The Virgo installation is in Santo stefano Macerata near Pisa, Italy. Like LIGO, Virgo is an interferometer that uses two 3-kilometre-long (1.8-mile-long) arms to detect gravitational waves. It was built by a collaboration of laboratories in Italy, France, the Netherlands, Poland, Hungary, and Spain. It is administered by the European Gravitational Observatory (EGO), and is named after the Virgo Cluster of galaxies, which lies some 50 million light-years from Earth.

Construction for Virgo began in 1996 and it was commissioned in 2003. It then did several data runs, and as with LIGO, it was upgraded to gain more sensitivity. Virgo joined with the advanced LIGO detectors and in June 2017, it was part of the first discovery of gravitational waves that LIGO detected.

Virgo and LIGO aren't the only gravitational wave interferometers in use on Earth. The Kamioka Gravitational Wave (KAGRA) detecter is nearing completion by the University of Tokyo in Japan. It has two 3-kilometre long arms, and uses lasers to performing interferometric gravitational wave detections. The facility is part of the Kamioka Observatory, where neutrino experiments are being performed. As with LIGO and Virgo, KAGRA is expected to be sensitive to gravitational waves from such events as binary neutron star mergers millions of light-years away. Upon completion in late 2019, KAGRA is expected to join with LIGO and Virgo in a series of joint gravitational wave observing campaigns.

A Quake Observatory on Mars

One of the more unusual space-based detectors in use today is sitting on the surface of Mars. It's called the Interior Exploration Using Seismic Investigations, Geodesy and Heat Transport (InSight) mission and landed on Mars on 26 November 2018. It was designed to study the interior of the Red Planet using seismic detectors to sense marsquakes and take the interior temperature.

Mars, like Earth, formed some 4.5 billion years ago and it is layered like Earth. It has a core, a mantle, and an outer crust. All of the missions to it had been concerned with the Martian surface and atmosphere. However, to truly understand a planet, it's important to know something about its interior construction. In particular, Mars poses some interesting questions because it is so Earth-like in many ways, yet totally alien in others. If, as many planetary scientists suspect, Mars was wet and warm early in its history, what happened to change it? The Mars MAVEN mission (mentioned in Chapter 5), along with India's Mars Orbiter Mission, is studying the atmosphere of Mars to understand how the atmosphere and some of Mars's water has largely dissipated to space.

Could the interior activity of Mars also be part of the answer to Mars becoming a dusty desert planet with water frozen underground and at the poles? InSight hopes to answer those questions by taking Mars's vital signs. Mission scientists describe taking the pulse of Mars and they do this by using the Seismic Experiment for Interior Structure (SEIS). It looks like a small dome deployed out from the lander and its sole job is to wait for seismic tremors caused by marsquakes and meteorite impacts. Other sensors on-board the lander measure wind speeds, temperature changes, and magnetic field fluctuations. The HP3 Heat Probe (nicknamed 'the Mole') is essentially a sensitive thermometer. It should eventually burrow about 5 metres (16 feet) down into the Martian surface to measure the interior heat from Mars. The data from the Mole will help determine what the central heat source is inside the planet. The RISE instrument keeps track of the exact location of the lander, almost like a super-sensitive GPS. This is important because Mars (as Earth does) wobbles on its axis. Information from this instrument, in addition to the SEIS detector, will help give a clearer picture of the Martian core. For example, no one is quite sure if the iron core is liquid, or if it is similar to Earth's core.

InSight also has a camera mounted on a robotic arm and has been sending back colour 3D images from surface via a radio antenna that relays data through spacecraft orbiting Mars, and from there back to Earth and the Deep Space Network. Another camera is used to provide wide-angle views of the area where the lander is working.

The lander and its 'mole' detected its first marsquake on 6 April 2019 when it picked up a very faint seismic signal. Data analysis was ongoing as of this writing, but if it is confirmed to be a quake, the signal very likely came from deep inside the planet. The signal was so faint that if the quake had happened on Earth, it would have been just one of many tiny tremors which ring through the planet each day.

MarCO CubeSats

InSight did not travel to Mars alone. It was accompanied by two small CubeSats called MarCO A and B, (for Mars Cube One), which were quickly nicknamed EVA and WALL-E. They were part of an experiment to supply a communications relay system for the mission. Each of these tiny satellites measure 10 cm (nearly 4 inches) square, equipped with solar panels, and radio receivers and transmitters. Once they got to Mars, they provided up-to-the-second information about the InSight mission as it went through the entry, descent, and landing on the planet. EVA and WALL-E were essentially on a fly-by mission, and are now in an orbit around the Sun. NASA's last communication with them was in late December 2018 and January 2019.

Searching for a Primordial Element Using the Deuterium array

In all the work that telescopes and observatories do, across all the wavelengths of light and the high-speed trips taken by neutrinos and cosmic rays, there's at least one cosmic mystery scientists haven't yet fully explored: the conditions in the very earliest epochs of the universe. Through images of the first stars and galaxies, astronomers get a good idea of when they formed, only a few hundred million years after the Big Bang. But, those images and other data still don't explain the existence and distribution of dark matter, or the phenomenon of dark energy. Those are mysteries which remain to be solved. However, there are tantalizing hints and clues, particularly when scientists look to an isotope of hydrogen called deuterium. Astronomers have known for quite a while that the amount of deuterium in the universe has a bearing on how much regular (baryonic) matter there is, along with light. They are related.

Deuterium can serve as a probe of conditions in and after the Big Bang, where nearly all of it was created. In addition, astronomers want to know how much of this isotope existed in the early universe and compare it to its abundance now. This knowledge helps them calculate the density of regular matter in the universe, and shed some light on how much ordinary matter is hidden away in black holes, gas clouds, brown dwarfs, and stars. Even though astronomers have a good idea of how much ordinary matter there is (about 5 per cent of the total mass of the universe), pinning down an exact amount also helps put some boundary numbers on the amount of dark matter and dark energy.

Nearly all deuterium was created in the Big Bang, although there are processes which create a small amount now. However, deuterium is more easily destroyed, particularly in the interiors of stars. A small amount of deuterium relative to hydrogen in a region means that most of the deuterium which was originally located there has been destroyed or locked into a chemical bond with another element. So, for example, the more star formation is in a region of a galaxy, the less deuterium it is likely to have.

Planetary scientists study the ratio of deuterium to hydrogen in comets, for example, trying to understand where Earth's water originated. Comets have long been thought to be one mechanism that delivered water to our planet in its infancy. The so-called 'D/H ratio' could help determine whether or not comets were involved. Comets contain water largely unchanged since they formed in the very earliest epochs of solar system history. Studies of cometary particles showed some comets had the same D/H ratio as Earth water. However, some comets have different ratios than Earth water, and may not have been viable sources.

It also turns out the amount of deuterium in the universe can be related to the amount of dark matter. Measurements of deuterium, however, are difficult to make, even with sensitive studies of the molecules of gas that exist in the interstellar medium. An accurate assessment of how much deuterium there is in the universe would let scientists refine their models of the Big Bang. It would also give scientists a much better idea of how much baryonic matter there is (which is the matter we can detect in the universe).

The big challenge in figuring out how much deuterium there is, and in what proportion to regular baryonic matter, is that atoms of this element are very hard to detect using Earth-based instruments. There's roughly one deuterium atom for every 100,000 hydrogen atoms. Doing optical spectroscopy to find it is difficult because the hydrogen spectral line is very close to the deuterium line. However, it turns out deuterium is easier to spot at radio wavelengths, but there's a hitch. The radio 'signal' which deuterium emits is at 327 MHz, which can easily be 'swamped' by something as simple as an answering machine, analogue TV sets, cell or mobile phones, and fluorescent lights in a nearby building. So, to search out deuterium in the interstellar medium requires a fairly radio-quiet sensing area.

This was the challenge facing scientists at MIT's Haystack Observatory when they built an array of receivers to search for the faint whisper of deuterium in our galaxy. The instrument, called the Deuterium Array, was a set of twenty-four stations tuned to the frequency of deuterium. It covered an area about the size of a soccer field. Each station looked like a dipole antenna. Before the array could go into operation, the science team had to search out

Aerial photograph of LOFAR radio telescope in 2011. Courtesy of Dr G.B. Gratton, STFC.

sources of radio frequency interference in the near neighbourhood. Since Haystack is in a fairly well-populated area outside Boston, Massachusetts, the team went around to all the neighbours and either replaced 'noisy' equipment or figured out ways to make stereo equipment and other sources of spurious signals quieter.

The array began searching for deuterium in a region of the Milky Way galaxy where they suspected significant amounts of primordial deuterium (dating back to the Big Bang) existed. After a year of operation, it successfully detected deuterium in the plane of the Milky Way galaxy. The experiment's design also served as a technical pathfinder for arrays such as LOFAR, SKA, and the Murchison Wide-field Array (MWA).

Are They Out There?

The search for extra-terrestrial life is astronomy's answer to the question: 'Are we alone?' While the idea of 'aliens' is a standard trope in science fiction, as well as popular with legions of UFO 'true believers', the truth is that we have no way of knowing right now if there *is* life out there. Exoplanet searches have certainly confirmed other worlds exist in our galaxy, and most likely in others, too. But, detecting life on those planets is still very difficult.

There are several ways to look for life. One is to look for signs of life in an exoplanet's atmosphere. Or, planetary scientists can send craft to look for signs on or beneath the surface of a world. Another way is to search for signals, since it's reasonable to assume that extra-terrestrial intelligence will use some form of electromagnetic radiation to send signals across space. It's is a big assumption, and some researchers have suggested a number of reasons why the method might not work. For one thing, ET (as people like to refer to alien life forms) may not be communicating using 'radio' or other methods. The aliens might not be that far advanced, or they might be so far advanced that they don't use 'old fashioned' signals the way we do. But, assuming they do, it's reasonable to think their communications could be detected. So, a number of experiments have been conducted over the years to look for the faint radio emissions that might signify intelligent life in the galaxy. People have even suggested looking for the redshifted signals of military radars at distant worlds!

The idea of using signals to find aliens isn't a new one. It stretches back as far as the late 1800s, when various people (including Nikola Tesla) wrote that it might be possible to contact beings on Mars using his wireless transmission systems. This idea was part of the general interest in Mars that swept over the world. Other experiments were performed using various receivers, but of course no signals from Mars were ever detected.

One of the earliest SETI searches took place in 1960 and was called 'Project Ozma'. It used a radio telescope at Green Bank, West Virginia, (the predecessor of the Robert C. Byrd Green Bank Telescope) to search for signals that might be coming from two stars: Tau Ceti and Epsilon Eridani. Both lie fairly close to the solar system, at distances of 12 and 10.5 light-years respectively. Scientist Frank Drake tuned the telescope to look at a frequency of 1,420 gigahertz. The significance of this frequency is not random: it's very close to the frequency of the hydrogen and hydroxyl radical lines. Scientists have dubbed this the 'water hole', since both are part of water, and water is assumed to be essential for life. It would be a natural frequency for civilizations to use to contact each other, like a gathering around a water hole in the desert.

Over the years, other such searches have been made, using Arecibo in Puerto Rico, Parkes radio telescope, and other facilities. The history of these searches is a fascinating one, involving many well-known scientists as well as political obstacles and funding problems. The SETI Institute in California was formed to continue looking for signals from outer space, and today there are international efforts that include LOFAR and MWA, as well as the Lovell Telescope and Arecibo.

The Allen Telescope Array (ATA) is another search effort spearheaded by the SETI Institute, with funding from a number of donors including the late Paul Allen, who co-founded Microsoft with Bill Gates and endowed the project with $30 million. Its 42-dish array, located at the Hat Creek Observatory in northern California, scans the skies, looking for signals which could be from distant civilizations. The idea behind it was to build a large array with a number of smaller dishes, and was first developed as a joint research project between the SETI

Institute in San Jose, California, and the University of California, in Berkeley. The array is now owned by SRI International, which manages Hat Creek Observatory, and operates the ATA. In the past, the array has studied the skies for signals, and in 2017, studied the strange interstellar asteroid 'Oumuamua. This object was discovered in October 2017 and was originally thought to be a comet. Its trajectory took it past the Sun and out of the solar system.

'Oumuamua's origin is from outside the solar system and that's why the ATA was tasked to study it. The idea was to look for any signs of technology on the asteroid, as if it could be a spacecraft. Unfortunately, there were no emissions from the asteroid.

How Big Can a Telescope Get?

Over the years, telescopes have become bigger and bigger as astronomers seek to look deeper and further into the universe, to study fainter objects in more detail. The current largest 'single-eye' facilities are the FAST radio telescope in China at 500 metres (1,640 feet), the largest single-dish radio observatory in the world, and the Extremely Large Telescope in Chile, at nearly 40 metres (131 feet) across. But, ELT had a predecessor that never got built: the Overwhelmingly Large telescope (OWL). It was first proposed in the late 1990s and would have used a segmented mirror design using a whopping 3,264 separate segments to create one large reflecting area. This is because building a single mirror is incredibly challenging not only to build, but also to transport to an observatory site, and to place it in a telescope mount. Using the segmented design would also allow for an advanced adaptive optics system to help correct the telescope and maintain its sharp view.

So, the OWL stayed on the drawing boards for many years because it would have allowed astronomers to study planets around Sun-like stars and analyse their atmospheres, as well as see objects too dim for even Hubble to see. Among them would have been the first stars to ever form, nearly 13.7 billion years ago. It's easy to see why astronomers would be excited about detecting those. However, while the design was amazing, and feasible, the cost was not – so, eventually, the OWL concept was abandoned. Instead, the European Southern Observatory opted to build the Extremely Large Telescope (ELT).

Even though the OWL lost out in the great race to build larger telescopes, its design challenges informed the design and creation of the ELT. Its forward-looking technology will be used in future large observatories.

Losing Interest in a Telescope

Observatories have come and gone throughout the history of astronomy. Our planet is littered with ancient, little-used observing sites from ancient times. Stonehenge in England and the Medicine Wheel in Wyoming come to mind. But there are others. Want to look at the site where Tycho Brahe built his instruments on the island of Hven? For a long time, it was abandoned, apparently lost to history. Then, in the 1950s, archaeologists excavated in the area and the observatory was restored to some of its glory. Today, there's a museum, but the facility is no longer an observatory. Fast-forward to the 21st century, and witness the closure of Yerkes Observatory. It became too costly to maintain and little research was being done there. Do a web search for 'observatories' and major lists of them appear, and a fair number of facilities are listed as 'defunct'. Most are older, or failed in some other way.

For example, the Meyer-Womble Observatory on Mount Evans, in Colorado, was commissioned in 1995 by the University of Denver at an elevation of 4,326 metres (14,192 feet) above sea level. At that rarefied height, the facility became the third-highest optical/infrared observatory in the world. While it was in service, unfortunately, it was severely damaged in a high wind storm in the winter of 2011/2012. The old facility had to be torn down, and a new one erected. Unfortunately, the cost of maintaining the observatory was more than the University wanted to shoulder, and the site has been abandoned (at the time of writing). The observatory is now slated for decommissioning.

This is not an unusual fate for a facility. Budget constraints, technical challenges, outdated equipment, light pollution, infrastructure issues and the race to keep instruments in good working order, can challenge observatory operators. Sometimes it's just easier to build a new facility rather than to keep trying to upgrade an existing one.

In other cases, one observatory is sacrificed so another may be built (either in the same spot, or nearby). The United Kingdom Infrared

Telescope (UKIRT), which was owned by the United Kingdom Science and Technology Facilities Council, is now funded by NASA and used by a consortium of universities and private industry. It was used to for the UKIRT Infrared Deep Sky Survey, which looked at a wide area of the northern hemisphere sky, at brown dwarfs, distant dusty starburst galaxies, distant galaxy clusters, and quasars in the very early universe.

As part of the negotiations for the placement of the Thirty Meter Telescope on Mauna Kea, astronomers had to agree to remove several facilities. UKIRT is one of them, and it will be decommissioned in the very near future. For now, however, it is still producing scientifically useful data.

Aurorae, Ionospheric Observatories, and Controversies

Most observers in the northern and southern latitudes of the world are familiar with the polar lights, commonly called aurora borealis and aurora australis (northern and southern lights). These occur when the solar wind slams into Earth's magnetosphere (the region of space influenced by our magnetic field). Energised particles in the solar wind get funnelled down the lines of Earth's magnetic field toward the poles and collide with atoms and molecules of gases in our atmosphere. The collisions create the aurorae. Different colours in the aurorae indicate different gases that are being excited.

Aurorae are one manifestation of solar activity and how it impacts our planet. Activity on the Sun is the source of space weather. Sunspots, coronal mass ejections, and flares are all outbursts which eventually send clouds of charged particles out through the solar system. Our solar-observing satellites sense these disturbances and give some early warning. And, our planet is not totally defenceless. Earth's magnetosphere protects us from these particles, along with the upper atmosphere. However, there are effects which can be observed from Earth, including the aurorae.

A number of facilities around the world study these solar wind interactions with our magnetosphere in an effort to understand the processes at work high above our heads. Haystack Observatory, for example, is heavily involved in making ionospheric observations throughout the year. It uses specific instruments for its studies, as well as signals from GPS receivers. The Millstone Hill Observatory at Haystack is what's called an 'incoherent scatter radar' instrument. It is used mainly to measure the scattering behaviour of electrons in Earth's ionosphere, at an altitude of 100 to 1,000 kilometres (62 to 621 miles). During space weather events, the density of electrons at those heights changes and the radar measures how dense they get and how quickly the density rises and then dissipates. It studies such changes over a territory that ranges from the mid-Atlantic out to the center of the US, all the way up to the Arctic Circle and down to Florida. It's critical information for understanding the planet's reaction to solar activity.

Another observatory focused exclusively on the behaviour of Earth's ionosphere, particularly when it's being disturbed by space weather events, is the High-frequency Active Auroral Research Programme (HAARPS). It was funded and built by the US Air Force and US Navy, with the University of Fairbanks, Alaska, and the Defense Advanced Research Projects Agency. It now belongs to the University of Fairbanks, and continues to study the ionosphere using the Ionospheric Research Instrument. It is basically a radio transmitter array that stretches across about 40 acres of land near Gakona, Alaska. When it's in operation, the array transmits a signal up into the ionosphere, which energizes a small area. The operation allows scientists to get information about how plasmas work in our upper atmospheric regions and what happens during interactions with the solar wind.

Upper atmospheric studies such as these, and those at Haystack, are important when it comes to understanding how the solar wind and solar outbursts can affect our planet. In addition, space weather generated by disturbances in the solar wind, can ultimately affect a wide range of technology we use for communications, power grids, and other infrastructure. Planes, trains, and barges, as well as GPS systems in our phones, cars, and other technology can be affected by space weather.

Interestingly, HAARP has drawn a lot of criticism by conspiracy theorists who think that (because of its former connection to the military) the array is really some kind of technology that can be used for bad purposes. Some of these people claim that HAARP could trigger floods, earthquakes, or even bombard people's minds with some kind of mood-altering rays. Of course, the array can do none of these things. It is merely a transmitter being used as a probe of the ionosphere so scientists can understand exactly what happens when the Sun sends a solar flare or a coronal mass ejection our way.

The really favourable implications of such research outweigh the fear-mongering by people ignorant of the real science being done at such places. In fact, the data which Haystack, HAARP, and other space weather observatories take can provide improvements to a wide range of technology. It's unfortunate that fear outsells fact, but in the long run, science usually wins and hysteria fades away. The scientific outcomes almost always outweigh any negative 'press' such a facility gets from those who don't understand its work.

7

OBSERVATORIES OF THE FUTURE

The science of astronomy has come a long way since the first stargazers stepped outside to use the sky. Over the centuries, humans have moved from being observers and users to become a species that seeks to understand what's 'out there' in the most fundamental ways we can. Observatories today offer a general glimpse of the most distant early reaches of time and space, but the details of some of those primordial events (and indeed, the Big Bang itself) still elude our vision. Where we once saw stars and nebulae, astronomers now also find planetary systems, but are still nailing down the details about how they formed.

The history of our own solar system is generally understood, but there are still questions about the early epochs of formation and the events that caused the protosolar nebula to coalesce and set off on the road to planet-building. Planetary scientists are still putting together an accurate picture of the links between planets and asteroids, and sources of water in the solar system. Astronomers see organic pre-biotic molecules in space, and are gaining a clearer understanding of how they formed. But, how did they create life? And, is there life elsewhere?

Physicists continue to pin down the characteristics of cosmic rays and neutrinos and sub-atomic particles, while astronomers use the information to ask: what is the nature of dark matter and dark energy? Those are among the many questions that face astronomers, cosmologists, astrobiologists, and planetary scientists in the future. To answer them, they need ever more sensitive telescopes and instruments to study objects and events in the cosmos. The ultimate goal is a deeper understanding of origin and evolution of the universe. That's where future instruments come in – both newly designed ones and updates to existing observatories.

The future of astronomy is both bright and challenging. There is still a lot to learn and observe. At the same time, there are fewer dark sky sites where astronomers can build new observatories. This means future facilities will need to be built in existing sites, and astronomers and the world's populations will need to figure out a way to mitigate the effects of modern civilization on all types of observatories. In some places, such as at Kitt Peak, light pollution is a growing factor, despite decades of trying to abate it in Tucson and other nearby towns. There are whole generations of students growing up not knowing what a truly dark sky site is, and the astronomers who get to visit the mountaintops and deserts are among the privileged few who can appreciate the velvety backdrop of night.

There are other threats to observatories which don't seem obvious at first. Funding is one - or, actually, the lack of it. These facilities are expensive to build and maintain. It's one reason why institutions and laboratories began collaborating on observatories. Shared costs are easier to bear for the universities, government institutions, and countries that want to build observatories. The same was done with observatories such as Hubble Space Telescope. International collaboration is now the key to building the large arrays and other facilities needed to do the 'Big Astronomy' of the 21st century.

While light pollution on Earth is a threat, orbiting satellites also affect the Earth-based view of the skies. The recently launched Starlink satellite constellation (deployed by SpaceX to enhance internet and other communications from space) threatens to impact science observations. The sheer number of these satellites, plus their brightnesses in both optical and radio frequencies, is affecting the work astronomers do. In particular, surveys for potentially threatening near-Earth asteroids could be made far more difficult than they already are. Solutions are being discussed that should help astronomers, but it seems clear that the skies as seen from Earth's surface have been irrevocably changed.

Optical observatories aren't the only ones being affected. Radio telescopes and arrays are seeking out the most radio-quiet places on the planet. Thankfully, there are still some where science can be done effectively. However, the widespread use of cellular technology and other electronics (even the electronic ignition on a vehicle can disrupt an observation), remains a threat to effective radio science.

Of course, the solution to light and satellite pollution and radio-frequency interference might seem obvious: send all our observatories to space. It's being done, of course, but space-based facilities are incredibly expensive and require extensive testing and hardening for the extreme environment. As long as humanity can afford to launch observatories to space, they will be sent.

In spite of the challenges, new observatories are on the drawing boards and under construction both for Earth- and space-based research. In addition, current facilities are planning upgrades, new instruments, and other enhancements to capitalise on the investments already made in them. It's time to polish up the good optical-quality crystal ball and take a look at what's coming for Big Astronomy.

Future Astronomy from Earth

Over the past few thousand years, astronomy has changed radically from simple observations to a huge data-gathering enterprise. Today, each of the world's major observatories delivers multiple terabytes of data daily. By one estimate, astronomers are swimming in at least 200 petabytes of information about the universe, and future observatories will deliver at least that much or more in a rapidly increasing manner. This ocean of data is already transforming our understanding of the cosmos, and has changed astronomers, as well. They are not just observers any more. They are becoming data managers and data sharers across the extensive collaborations they've formed. This is also transforming data collection, communication, and storage standards to ensure access to this hard-won knowledge.

The Extremely Large Telescope

The 21st century is rapidly becoming the Age of Giant Ground-based Telescopes. There are several large ones coming online in the next few years. Such behemoths are among the highest priorities for doing cutting-edge ground-based astronomy. Among them is the Extremely Large Telescope, which is currently under construction by the European Southern Observatory on Cerro Amazones in Chile's Atacama Desert. When completed, it will be the world's largest telescope sensitive to optical and near-infrared light. It will use a 39-metre (127-foot) primary mirror made of 798 segments. Each segment is 1.4 metres (4.5 feet) across and will be correctly positioned with actuators. These will help the mirrors focus light the same as one large mirror of the same size to the single-piece secondary mirror. In addition, the telescope will use two additional mirrors to deliver high-resolution final images to the detectors. Preliminary cost of the project was estimated at $1.34 billion (in 2012 dollars).

When ELT has its 'first light' observations, sometime in 2025, it is expected to do observations across a wide range of subjects. The search for planets around other stars is a major impetus for the telescope. Thousands of such planets exist, and astronomers know something about them thanks to observations by the Kepler mission, Spitzer and Hubble Space Telescope, and other projects. The next steps are to identify Earth-type worlds, and in particular, to study their atmospheres. The gaseous envelopes surrounding those planets hold clues to whether or not such places have life or can sustain life. Do they have water? If they have life, what sort of life is it? Ultimately, astronomers want to find other planets like our own, exo-Earths that may have formed in much the same way as our own planetary home.

Looking back across time and space, astronomers will use the ELT to search for the very first objects and events in the universe. Already, there have been tantalizing observations of very early stars and galaxies, but when did they start to form? There is a hard limit right now to how far we can 'see back' in time. It's called the Cosmic Dark Ages, and it was a time that lasted about 400,000 years after the Big Bang, and the very young universe was extremely dense and hot. Light and matter were closely coupled, and photons could not travel very far. Eventually, the rapidly expanding universe cooled enough so that hydrogen could form and light could travel freely. This is the point in time where scientists want to start looking for the seeds of stellar formation and the births of the first galaxies.

In addition, many questions remain about how the first galaxies formed and evolved. When did the first galaxies form? What kinds of stars do they form throughout their lives? How did the galaxies themselves change? The answers lie within the galaxies themselves, in the very stars and nebulae that exist within them. Astronomers need to look at individual stars in distant galaxies, and to study the structures of the earliest ones.

Finally, there's the question of dark matter and dark energy: what are they? Astronomers know they both exist, but their nature remains elusive. Gaining an understanding of both is fundamental to understanding what our universe is made of. We know that the matter we see – the baryonic matter – is only a very small fraction of the total mass of the universe. Another 22 per cent is dark matter, and the rest is dark energy. The ELT will be used to chart the distribution of dark matter throughout the universe, and explore the effect of both dark matter and dark energy on the stars and galaxies.

The Giant Magellan Telescope

Another member of the incoming class of giant telescopes is also currently under construction in Chile, at the Las Campanas Observatory up at 2,550 metres (8,366 feet). Called the Giant Magellan Telescope, it too will tackle some of the same questions that ELT and other telescopes want to answer. The GMT should begin operations in 2025, taking advantage of the dry climate and spectacular dark-sky views which have brought other telescopes to this site in South America. It's being created by a partnership of research and university institutions in Australia, Brazil, Chile and South Korea, led by the United States.

The mirror at the heart of GMT is a segmented one consisting of seven smaller mirrors installed inside a cell. They will work together to create a 24.5-metre (80.3-foot) optical surface. Light will reflect from that larger surface onto seven smaller secondary mirrors, and then to a CCD detector. The telescope's designers claim it will have a total collecting area of 368 square metres and have a resolving power ten times greater than Hubble Space Telescope's already sharp view of the universe. The telescope will employ adaptive optics (as most modern large observatories already do) which will use actuators under the secondary mirrors to shape them and allow them to counter the effects of atmospheric turbulence. When needed, a laser guide-star system will 'sample' the atmosphere and feed data to the system so it can correct for any aberration. The light it gathers will be fed to a series of imagers, near-infrared and visible light spectrographs, and specialised cameras.

Once it has been commissioned, the GMT will make observations of distant planetary systems, nebulae, galaxies, and the large-scale structure of the universe. Its designers also want to tackle one of the most evocative questions that astronomers can ask: 'Are we alone?' Specialised instrumentation on the GMT which will allow it to detect the presence of Earth-mass worlds orbiting Sun-like stars in nearby space and gaze out to the limits of the observable universe to look for the first stars to ever form. The telescope will also continue in-depth studies of stars to understand the origins of some of the chemical elements which make up everything from planets and stars to living organisms. Finally, it will also make observations in hopes of solving the mystery of dark matter and dark energy – a mystery that will take all of the world's telescopes to uncover. As with other telescopes, it will look for evidence of the first stars and their formation, and the subsequent birth and evolution of galaxies.

The Large-Scale Synoptic Survey Telescope

Cerro Pachón, home of the Gemini Telescope, will soon welcome a new neighbour – the Large Scale Synoptic Survey Telescope (LSST). It is currently under construction on El Peñón peak, and astronomers expect this observatory to come online sometime in 2020. It will consist of a three-mirror, three-lens telescope and the world's largest camera, a 3.2-gigapixel CCD. The main mirror is an 8.4-metre (27.5-foot) reflector, with a secondary 3.5-metre (11.4-foot) mirror and a 4.0-metre (13-foot) tertiary. Together, they work to focus incoming light from distant objects into the camera detector.

Astronomers have some impressive goals for LSST. Its major 'mission' will be a ten-year-long survey of the sky which will provide in excess of 500 petabytes of images and data showing the large-scale structure of the universe. Large-scale structure is basically how galaxies and galaxy clusters are arrayed in a cosmic web of matter. Understanding this structure will also give insight into the distribution of dark matter and, by measuring the expansion of the universe, also create more information to help them understanding dark energy.

LSST will be measuring a phenomenon called weak gravitational lensing. Many readers have probably seen images of strong gravitational lensing where a strong gravitational source produces a warped set of images of a distant galaxy or quasar. Its light is 'bent' or warped as it passes by a strong gravitational field. There are galaxy clusters that work perfectly as strong gravitational lenses. But, they don't exist all over the sky. However, dark matter does exist everywhere, and it produces a weak lensing effect on the light from a distant galaxy or quasar. By measuring the redshift of distant galaxies, studying the light from distant supernovae and other phenomena, LSST will provide more data on cosmic expansion and possibly how the mysterious phenomenon of dark energy is affecting the expansion rate of the universe.

As it surveys the sky, LSST will detect novae, supernovae, gamma-ray bursts, and take measurements of the variable brightnesses of distant quasars. Its data will be used to notify other 'watch' surveys tuned into those types of objects. This telescope will also produce images and gather data about objects in the solar system. Of particular interest will be the locations and distribution of near-Earth asteroids – which can pose a threat to our planet if they stray too close in their orbital paths. In addition, LSST should be perfectly poised to make discovery images of Kuiper Belt objects. Finally, it should be sensitive enough to be used to detect the events and objects that cause gravitational waves.

Over its ten-year-long survey, the LSST will be able to cover 18,000 square degrees of southern hemisphere skies, making a total of more than 800 visits to each spot in the sky. All of the telescope's data will be made available almost immediately, with up to 15 terabytes per night uploaded to Google as part of the maps in Google Sky.

The Thirty Meter International Observatory

One of the newest of the large telescopes still in planning is the Thirty Meter Telescope (TMT), planned for Mauna Kea on the Big Island of Hawaii. It is an international facility managed by a consortium of astronomers from the United States, Japan, China, India, and Canada. Because Mauna Kea has some of the best observing conditions on the planet, and because it is already home to a number of optical- and infrared-sensitive telescopes, the site was chosen in 2009 after a survey of a number of sites around the world. Almost immediately, the selection was criticised by native Hawaiians who felt it was trespassing on what they consider to be a sacred mountain. The controversy had been brewing for years, as other telescopes were built on the summit. After nearly a decade of meetings, negotiations, court cases, blockades, and other problems, the observatory site was approved for construction. In return for building TMT, the astronomy reserve must divest itself of at least two telescopes, and TMT will be the last major telescope built on the site. There are other concessions that the telescope management group must make. The TMT has agreed to fund $1 million per year for science, technology, engineering and maths education on the island, and is working with local education leaders on programmes to use that money. In addition, the TMT leadership has created a Workforce Pipeline Programme to prepare Hawaii Island students for science and technology jobs.

The telescope at the heart of the TMT will have a 30-metre (98-foot) primary mirror, a smaller secondary mirror, and a third one to direct incoming light to any one of eight instruments attached to the telescope. As with other telescopes on the mountain, the TMT will use adaptive optics to correct for turbulent atmospheric conditions. The whole assembly will be a computer-controlled structure, which will allow almost unlimited motion for the telescope to point at wherever it needs to for all its observations.

As one of the largest ground-based observatories, TMT will perform spectroscopic exploration of the earliest epochs of cosmic history, when the first sources of light and the first heavy elements in the universe formed. The early universe is still a largely unknown territory for study, largely because it is so difficult to access. Objects at that time appear dim. Early galaxies appear as they did at the beginning of their formation and evolution. These challenges make telescopes such as TMT all the more important to build, to cast our gaze back to the birth of the universe we can detect.

TMT will also investigate the formation of black holes ranging across all of cosmic history. In particular, the supermassive monsters at the hearts of galaxies seem to play a role in galaxy evolution. What is this role? It's a question astronomers want to answer with TMT and other next-generation telescopes. Closer to home, TMT will study galaxies in the 'current' universe – the Milky Way and nearby cluster neighbours. The stars of the Milky Way and its companions function almost as a 'fossil record' of the billions of years of galaxy formation and evolution they have experienced. The numbers of stars, their distribution throughout a galaxy, and their ages provide crucial information about how stars and galaxies are born and are shaped over time. What astronomers learn from our own nearby neighbours helps them understand the same astrophysical processes that shape other, more distant galaxies.

Finally, planet-forming processes are an important part of the astronomy that TMT will investigate. In particular, the deaths of stars contribute to the formation of elements which get recycled into new generations of stars and planets. How star formation gets started, how it influences planetary accretion around different types of stars is a central area of study cutting across many disciplines in astronomy. TMT will complement the current generation of telescopes and carry out highly detailed investigations of young stars, dying stars, and the clouds of gas and dust where new planets are born.

The TMT should be ready for its first observations beginning in 2027, part of the next wave of 'the largest telescopes on Earth'.

Interestingly, astronomers are already referring to the present wave of 8- to 10-metre class scopes as 'medium size', in comparison to these new larger behemoths due to come online in the next decades.

A Next-Generation Gamma-ray Array

Gamma-rays, as we know from the previous chapter, don't usually make it all the way to Earth's surface. As they zoom through our atmosphere, they collide with molecules of gas. That produces cascades of ultra-high-energy particles, causing a characteristic glow of blue light called Cherenkov radiation. It's fairly faint; the human eye can't really pick it up. However, specific detectors can sense this light. The Cherenkov Telescope Array is being built to detect this radiation caused by gamma-ray collisions and turn it into data that will help scientists understand the sources of high-energy particles in the universe. The array is being built by a thirty-one country consortium that includes nearly 1,500 scientists and technical support team members.

The array will eventually consist of more than a hundred telescopes installed at locations in the northern and southern hemispheres. Together, they will comprise a huge collecting area some one million square metres in extent.

The science goals for CTA are impressive and represent a major new way to use Cherenkov radiation to study the gamma-ray environment of near and deep space. If it all works as planned, the array will probe the universe for such cosmic oddities as annihilating dark matter particles. It will also measure the extent of particle accelerating events that release gamma-rays and other high-energy emissions. Its primary targets will be center of our own Milky Way Galaxy, a region which contains the supermassive black hole called Sagittarius A*, the plane of our galaxy and sources of variable and transient behaviour among the stars, the nearby Large Magellanic Cloud (a satellite galaxy of the Milky Way), distant galaxy clusters and the active hearts of galaxies, plus highly energetic events and processes in the universe that release and accelerate particles across the cosmos. In addition, the array will search out star-forming regions to understand how and when cosmic rays affect the birth processes of stars. The Cherenkov Telescope Array should come online in the next few years. Construction began in 2019 and prototypes of the telescopes, detectors, and cameras are being tested.

Building The World's Largest Radio Telescope Array

Just as optical and infrared astronomers are building some of the world's largest 'glass' telescopes, radio astronomers are using arrays to create ever-larger detectors. These arrays, such as the VLA, create viewing areas as large as the extent of the array. The largest under construction at this time is the Square Kilometre Array (SKA). It's an international project being created by institutions in a dozen countries which will eventually deploy thousands of radio dishes and low-frequency antenna 'farms' to monitor the sky.

The SKA installation is two-fold. Part of it is being built in South Africa in the Karoo region, and the other part is in Western Australia in an area called Murchison Shire. The high-frequency detectors will be in South Africa, where a 64-dish array called MeerKAT already exists. The SKA collaborative will add another 133 antennas to the collecting area. The low-frequency arrays will be in Australia. Currently, a test project is being built out in Australia, near the site for SKA. It's called the Murchison Wide-field Array. Current plans call for 512 low-frequency stations to be installed in a spiral design across the Murchison area. More dishes and antennas will be added to each site as time and budget allow. The array is slated to start taking measurements of the radio skies starting in the 2020s, when the installation is only partially completed.

SKA is designed to answer those really big questions other observatories want to answer in different wavelengths as well. For example, how were the first stars and black holes created? The answer to that lies in the very earliest moments of cosmic time, back before light could travel freely through the infant universe. This was only a few hundred thousand years after the Big Bang. What could SKA possibly find there? As it turns out, there's plenty. SKA can focus on very weak signals from the neutral hydrogen that existed at the time. Those signals will be redshifted (that is, its wavelengths are shifted toward the red part of the spectrum), and SKA should be able to see them before, during, and after the epoch of reionisation, when light was able to travel across the rapidly expanding young universe. The distribution of those signals will also tell astronomers something about when and where the first luminous sources (stars and galaxies) began to shine. The investigations SKA will make in this area of study can only be done at radio frequencies, and will complement the already extensive optical and infrared observations stretching back almost to the same time.

Another question SKA will tackle focuses on magnetic fields in space. What generates them? Astronomers know they exist, and that they have an effect on star-forming regions, galaxies, and other regions in space. However, magnetic fields aren't visible by themselves. They can be inferred by other activities and objects. A black hole, for example, may have a jet emanating away from the disk of superheated gas surrounding it. The jet may be emitting synchrotron radiation, which indicates the presence of a magnetic field entrained in the jet. Such jets give off radio frequencies, which can be detected from here on Earth. SKA will use several methods to measure for magnetic fields in distant galaxies and clouds of interstellar gas. Those studies will help astronomers map the magnetic fields in our own galaxy, and understand the characteristics in fields of other galaxies.

SKA will also look at galaxy evolution, the role of dark matter in the universe, as well as try to find an answer to the question: what is dark energy? Just as the telescope array will study neutral hydrogen from the earliest epochs after the Big Bang, it will also look at the role of the gas in very young galaxies in the infant universe. It will track how galaxies rotate and measure the motion of hydrogen gas as they do. This will give another probe of the amount of dark matter in evolving galaxies. Sometimes the gas rotates at a different rate from the rest of the galaxy, and dark matter may play a role in constraining those rates.

Finally, astronomers using the array will attempt to answer the question: 'Are we alone?' That query stands at the heart of so much astronomy, as scientists search for the pre-biotic chemicals which make up life. In addition, the telescopes can be used to 'tune in' to any possible signals from extra-terrestrial civilisations who may be reaching out to see if they, too, are alone.

Developing a Future Neutrino Detector

As we saw in Chapter 6, neutrino observatories and detectors are tapping into a hot area of study which combines astronomy, astrophysics, and particle physics. While there are a number of detectors already online, more are on the way. One of the most unusual is actually located deep under the Mediterranean Sea, secluded away from errant cosmic rays. It's called KM3NeT, which stands for Cubic Kilometre Neutrino Telescope, and, when fully built, will include installations off the coasts of France, Greece, and Sicily. The design for this set of detectors is based on several other projects: the ANTARES detector (also under the Mediterranean Sea), NEMO (the Neutrino Mediterranean Observatory), and the NESTOR (Neutrino Extended Submarine Telescope with Oceanographic Research) project. ANTARES (which stands for Astronomy with a Neutrino Telescope and Abyss Environmental Research) is a project that was deployed 2.5 kilometres (1.5 miles) under the surface of the Mediterranean beginning in 2006. NEMO and NESTOR were under development, but were never built.

Construction began in 2012 with some prototypes and full strings of detectors deployed. At the time of writing, two sets of units are in use and connected by a high-bandwidth seafloor network. The data from the detections travels to a computer 'farm' onshore, and then to a central data collection and storage point.

It's an array which will, when it's completed, have around 12,000 spheres attached to strings of 18 spheres each. The spheres, called 'digital optical modules', are filled with photomultiplier tubes capable of detecting the radiation given off when a neutrino hits the sea water.

The goals for the KM3NeT are fairly simple: to discover and observe neutrino sources (such as supernovae and gamma-ray bursts) in the universe and then to determine the masses of the particles as they speed through the water. When fully built, this array will also work with the IceCube project to do wide-area neutrino detections from cosmic sources.

Future Space Telescopes

Future observatories in space are well along in development and planning. The Russian space agency will (as of this writing) be launching a satellite called Spektr-RG, which is a joint mission with Germany to do high-energy astrophysics in x-rays. A four-year survey

The main mirror of the James Webb Space Telescope is made up of hexagonal mirrors and is foldable for launch. After deployment, the mirror will unfold. Courtesy of NASA/James Webb Space Telescope.

of the x-ray sky is planned which will focus on galaxy clusters and high-energy activities in galaxies with supermassive black holes. Its observations may also help in understanding dark energy.

Structuring the Webb

Perhaps the most ambitious and closest to launch is the James Webb Space Telescope. It's supposed to be a follow-on mission to the venerable Hubble Space Telescope, although it may well be that they will overlap in operations. Whether it happens or not depends on how long HST will continue to work and when JWST will finally be launched.

The JWST is built to study the universe in infrared light. The wavelength range it's sensitive to reaches from long-wavelength visible light through the mid-infrared. This is different from Hubble, which is sensitive to optical, ultraviolet and infrared. Webb's science objectives are ambitious. For example, it will be able to look out to the earliest epochs of history to see some of the first stars and galaxies. These show up in infrared wavelengths, even though the original emissions were likely in ultraviolet and optical light. This is because the universe is expanding. As light from more distant objects travels, it gets redshifted (its wavelengths are shifted toward the red part of the spectrum). Those objects and events at very large distances look very redshifted, and easier to see in the infrared range. The Webb will be very sensitive to that light, which means the telescope will be able to 'look back' to a point in time just a few hundred million years after the Big Bang. This is when the first stars began to shine, followed by the formation of the first galaxies.

JWST will also be able to study the processes of star birth due to a unique property of infrared light: it can pass right through clouds of dust that scatter visible light. It allows the light from new-born stars to be detected in infrared when optical telescopes can't see anything due to the clouds obscuring the event. Likewise, it will be able to pierce the veil of gas and dust surrounding the cores of active galaxies, and the events leading up to the formation of planets in circumstellar disks. Finally, JWST will play a role in observations of objects within our solar system. Unlike stars, planets, asteroids, dwarf planets, comets, and even interstellar dust clouds can be more easily studied by Webb, using a mid-infrared detector.

JWST, which is named for former NASA administrator in the 1960s, is a NASA mission that is also funded by the European Space Agency and the Canadian Space Agency. It is expected to be launched and deployed in 2021 aboard an Ariane 5 ECA rocket from the Kourou ELA-3 site. The mission has faced redesigns, funding issues, and some damage during testing, but has recently passed most of its pre-flight vacuum and other testing.

How will Webb do its job? The Webb is equipped with a segmented mirror that will, when deployed, stretch across 6.5 metres (21 feet). To keep the telescope from overheating, the mirror will be shielded from sunlight by a large sunshield. The telescope will not orbit around Earth, which would interfere with its operations. Instead, JWST will be directed to a solar orbit about 1.5 million kilometres (93 million miles) away from our planet. Mission operations will be directed from the Space Telescope Science Institute in Baltimore, Maryland – the same organisation currently directing Hubble Space Telescope.

Unlike Hubble, the JWST is not serviceable, due to its planned orbit. So, the telescope has to rely on its own on-board resources, such as instrument coolants and fuel, to keep it in the correct orbit. The mission should last at least five years, but will have enough resources to stay in orbit for ten.

Studying Planets

Planetary science is also on the mission docket for the next decade, starting with the Mars 2020 mission. The spacecraft is under construction at NASA Jet Propulsion Laboratory in Pasadena, California, and launch is scheduled for mid-2020. The planned landing site for the mission is called Jezero Crater, in a region of Mars that is thought likely to have harboured a lake in the ancient past. It appears to be rich in clays, which can only form in the presence of water. Such a region could have supported life billions of years ago, which is one reason why the lander will settle down there for a long-term study. The search for life on Mars is an important driver as scientists plan for future human missions there.

Once landed, Mars 2020 will collect rock and surface soil samples for later pickup and delivery back to Earth by another mission. Since the mission is a rover, it will be able to cache a number of samples across its study area. Eventually, another mission will be sent (one still under study and design) to retrieve the samples.

Understanding the presence of water on Mars, both now and in the past, is a recurring theme in exploration of the Red Planet. Not only will it help planetary scientists understand the warm wet past of the planet (and what happened to much of the water), but ongoing missions also can identify currently existing water that could be of use to future crews landing for long-term human study of Mars.

The Mars 2020 rover will carry multiple cameras, including a SuperCam to do imaging, chemical composition analysis, and study the mineralogy of rocks in the Jezero Crater region. It will also have a stereo Mastcam similar to the one on Mars Curiosity. Other instruments will study the composition of Martian surface materials, a set of sensors which will monitor near-surface temperature and weather conditions, a drone, microphones (since no one has captured sound on the planet before) and other experiments designed to study the atmosphere.

Long-Term Future Prospects

It takes a long time to plan and build a space telescope, as we've seen with Hubble Space Telescope, the other Great Observatories, and the James Webb Space Telescope. In the US, such mission planning is often informed by what's called a 'decadal survey', in which astronomers poll the community about the next generations of telescopes to be built. This survey is a ten-year plan that summarises input from scientists in the US and beyond.

The European Space Agency has also identified the missions it would like to do over the next decade or so, in a plan called Cosmic Vision 2015–2025. Many are destined for space, which is always a high-risk environment.

China has a plan for a giant space telescope called Xuntian that it wants to orbit around a future space station. It would have a 2-metre (6.5 foot) primary mirror and a suite of instruments for planetary studies and astrophysics.

Japan wants to deploy advanced satellites for radar mapping, Earth observations, and solar corona studies. Currently, it is collaborating with NASA, the European Space Agency, and other agencies on the X-ray Imaging and Spectroscopy Mission (XRISM) to create a telescope to replace one called Hitomi, which failed not long after launch in 2016. The spacecraft will carry two instruments sensitive to so-called 'soft' x-ray emissions from such energetic objects and events as active galactic nuclei, as well as processes in galaxy clusters.

India has ambitious plans for solar telescope deployment as well as imaging satellites, and a mission to Venus. There's no shortage of space missions for astronomy, astrophysics, and planetary science.

WFIRST

NASA's Wide-field Infrared Survey Telescope (WFIRST) has been in development since the US National Research Council first made it a priority in a decadal survey in 2010. Six years later, the mission was approved to move ahead. It will loft a 2.4-metre (~8-foot) telescope to space that will take infrared observations of the universe using a wide-field 288-megapixel near-infrared camera, and a coronagraphic instrument that combines a spectrometer and camera.

WFIRST has faced a rocky road to construction, mainly due to presidential insistence it be cancelled. It has been saved for the moment, and its optical assembly contract has been issued. Launch of WFIRST is slated for the mid-2020s and once it is placed in space, it will migrate out to a stable orbit around the Sun to start a five-year mission. The science objectives for WFIRST are to search for exoplanets, do direct imaging of them using the coronagraph to block starlight from the parent stars, look for an answer to the question of what dark energy is by studying the fluctuations in the density of the matter we can detector (baryonic matter), observing distant supernovae and doing observations of weak gravitational lensing.

The WFIRST project is a collaborative effort involving NASA, the National Center for Space Studies in France, the German Aerospace Center, the European Space Agency (which is developing its Euclid project to do similar observations), and the Japanese Aerospace Exploration Agency. The total cost of the project is expected to top $2 billion dollars by the launch date, provided there are no delays or other problems.

Euclid

ESA's Euclid Mission is Europe's mission to map the geometry of the so-called 'Dark Universe'. The idea is to understand the role dark energy has played over the history of the universe and the role that dark matter has played. To do this, the spacecraft will use a specialised telescope to measure the shapes of galaxies at varying distances in the cosmos while studying the relationship between distance and redshift. Since dark energy has apparently been causing the universe's expansion rate to accelerate, the main role of the mission will be to find out why this is happening and how it may control future cosmic evolution.

The heart of the Euclid telescope will be a 1.2-metre (4-foot) mirror that will send light to a panoramic visible-light imager, and a near-infrared photometer and spectrograph. These will help astronomers probe the expansion history of the universe and trace the development of large-scale structure. Like WFIRST and JWST, Euclid will travel to a Lagrange point orbit, where it will do a planned six-year observation run. During this time, it will observe at least 15,000 square degrees of sky. It will take three 'deep field' survey images as well, and the entire mission will supply hundreds of thousands of images and dozens of petabytes of data for further analysis. Launch is planned for some time in 2022.

Protecting Earth

Closer to Earth, the European Space Agency has proposed the Hera mission, which is planned to rendezvous with a binary asteroid system called Didymos. It's classified as a potentially hazardous asteroid that could pose a threat to Earth if its orbit changes to intercept our planet. NASA is planning a mission called the Double Asteroid Redirection Test (DART), which would intercept the asteroid and its companion and send an impactor into the surface. Hera would follow-up and assess what changes the impactor makes (if any) to the asteroid's motion and orbit. DART will be launched in December 2020, and Hera is planned for launch in 2023.

The interest in such near-Earth asteroids is important. It's well known that many smaller worlds such as Didymos and its companion follow orbits which bring them close to Earth's orbit. Any perturbations in their paths could bring those objects to potential collisions with our planet. These worlds are tough to spot due to their sizes and dark colour. Space agencies have launched several surveys to spot potential impactors. If one is seen early enough and does pose a threat to our planet, then – given enough lead time and technology – it could be possible to find a way to nudge them out of their orbits, destroy them, or find some other way to avoid a potential catastrophic collision with Earth.

The Next Decade and Beyond

Observatories to be built in the late 2020s and the 2030s are largely the subject of decadal surveys being done now among the world's scientists. It's clear that a number of research directions will still be 'hot topics' a decade or more from now. Among them will be high-energy astrophysics, dark matter and dark energy studies, and a continuation of the search for exoplanets, particularly Earth-like ones.

The European Space Agency is looking to continue its work in two regimes: one as a study of exoplanet formation and the other a study of the hot, energetic universe. The first spacecraft, slated for a 2028 launch, is called the Atmospheric Remote-sensing Infrared Exoplanet Large-survey (ARIEL). As part of the Cosmic Visions plan, this observatory will study exoplanetary systems in an effort to understand the conditions for planetary formation and the processes that lead to worlds being built up from the dusty disks around stars. ARIEL will observe about a thousand exoplanets, mostly super-Earths, gas giants, and super-Neptunes circling different types of stars. The on-board instruments will be spectrometers to study the atmospheres of these worlds by studying starlight as it passes through those gas envelopes. They will look for various atmospheric gases (water vapour, carbon dioxide, methane, ammonia, and others), and provide some idea of what conditions existed when these worlds formed. The end result will be a catalogue of planetary spectra, an in-depth picture of the chemical nature of these classes of exoplanets.

ARIEL is planned to complement the planet searches of NASA's Transiting Exoplanet Survey Satellite (TESS) and other past and planned missions that have catalogued and surveyed worlds around other stars.

In addition to ARIEL, the European Space Agency is in the throes of planning the Advanced Telescope for High Energy Astrophysics (ATHENA). This proposed five-year mission would launch around 2031 and chase answers to the following questions: how does ordinary matter assemble into the large-scale structures (galaxy clusters and superclusters, and the larger cosmic web), and how do black holes grow? What is their role in shaping galaxies and the larger universe? The telescope will use instruments to map hot gas in the universe, particularly in galaxy clusters and the space between galaxies. Of particular interest is tracing how the gas accumulations evolve over time. What are their sources? The questions about black holes will require imaging and study of the supermassive behemoths at the hearts of many galaxies. There are many questions about these monsters: how old are they? Were they in the early universe? If so, what role did they play in shaping galaxies we see today? In addition, the telescope will track how black holes consume matter and how nearby material manages to escape the voracious gravitational appetites they have.

ATHENA will have a large x-ray telescope as well as a wide-field imager, and the entire spacecraft will be launched into a Lagrange point orbit. From there, it will perform about 300 planned observation programmes a year, plus any 'target of opportunity' observations of gamma-ray bursters and other energetic events.

In the United States, the most recent decadal survey for both planetary exploration and astrophysics is underway and will be concluded and released in 2020. Already, astronomers have identified several important missions they would like to see happen in the next ten-plus years. The first is the Habitable Exoplanet Observatory (HabEx), which is still largely in the concept development stage. It is slated to directly image worlds orbiting sun-like stars, including gas giants and ice giants, but mainly looking for Earth-like worlds. It will, like ATHENA, study their atmospheres in a search for signs of habitability: the presence of water vapour, oxygen, or ozone. In addition to planetary studies, the spacecraft will also make observations of the early universe and study the origins and evolution of very massive stars. While all stars ultimately create and send elements to surrounding space when they age and die, the most massive ones are the source of most of the heaviest naturally occurring elements that are recycled into other stars, planets, and life.

HabEx will orbit at a Lagrange point, have a 4-metre (13-foot) telescope at its heart, and send light to a camera and an ultraviolet spectrograph (which will help it detect atmospheric gases). The spacecraft 'bus' will use thrusters to stay in a stable orbit and slew around to study different parts of the sky. The planned launch date is in the 2030s and the entire project is at the mission concept stage.

The Next HST-Type Telescope

Another observatory which could be selected is called the Large Ultraviolet Optical Infrared Surveyor (LUVOIR). As its name implies, this mission covers a much larger range of light and is intended to be a follow-up to the Hubble Space Telescope, which will likely be offline by the middle of the next decade. Given the wide range of light it will study, LUVOIR will be used to study everything from worlds of our own solar system, to extrasolar planets, out to the most distant reaches of the observable universe. It will do it in higher resolution and with next-generation instruments, which will give sharper views and more in-depth science insight into the structure of the universe and the evolution of stars and galaxies. In short, it will do everything that Hubble can do now, but better.

This telescope, often referred to as 'next-generation HST' has two different designs at the moment. The first design uses a 15-metre (50-foot) diameter primary telescope aperture, which will capture incoming light and feed it to four serviceable instruments. The other design suggests an 8-metre (26-foot) telescope and three instruments. The telescope is planned to be serviceable, despite its distant orbit. Both designs look very similar to the James Webb Space Telescope, and would essentially unfold themselves once they get to space.

The instruments planned for LUVOIR are the Extreme Coronagraph for Living Planetary Systems (ECLIPS), the High Definition Imager (HDI), the LUVOIR Ultraviolet Multi-image Spectrograph (LUMOS), and an ultraviolet spectrophotopolarimeter contributed by a consortium of ten European Research institutions. It's called POLLUX. The ECLIPS instrument will suppress the glare from stars so the telescope can spot any orbiting planets. It will be the first instrument to enable direct imaging and spectroscopy of Earth-like exoplanets. The HDI instrument will supply the same types of 'pretty pictures' which made Hubble Space Telescope so popular. It will be sensitive to wavelengths between near-infrared and near-ultraviolet. LUMOS will allow spectroscopy of multiple objects in UV and visible light. Essentially, it is the grandchild of the Space Telescope Imaging Spectrograph currently on-board HST. POLLUX will do ultraviolet spectroscopy as well.

LUVOIR is still under study, but if it is given the go-ahead, launch could be in the late 2030s. It could be taken to space aboard NASA's Space Launch System, or, if a second alternative is chosen, the spacecraft would launch aboard a New Glenn by Blue Origins (named after pioneering astronaut John Glenn, a New Glenn is a single configuration heavy-lift launch vehicle capable of carrying people and payloads) or a SpaceX Starship rocket. Like its predecessors HST and JWST, LUVOIR should be able to do deep-field imaging and spectroscopy. That would give very detailed views of the most distant objects in the universe, and provide insight into their motions, physical characteristics, and evolution.

Lynx to Space

The 2020 Decadal Survey by NASA identified high-energy astrophysics as an important area of continuing study. This includes follow-up x-ray observatories, which probe the most active regions and events in space. The Lynx mission is a planned satellite observatory that would bring x-ray astronomy to orbit again. In particular, it will also follow up on studies made by JWST, WFIRST, and the largest ground-based telescopes. It will observe what they detected, but in x-ray wavelengths. Lynx will follow in the footsteps of the ATHENA mission, with ten times better resolution than any other space-based x-ray observatory.

The three science 'pillars' that Lynx's planners envision are: black hole dawn, engines of change, and the hidden light of suns. Black holes are well-known to astronomers now, although there was a time when they were only thought to be theoretical. That changed with Hubble Space Telescope and other observatories which spied them out and measured their effects. Now, Lynx wants to find out how black holes got started. In particular, scientists want to probe the link between the births of supermassive black holes and the galaxies where they are embedded. Interestingly, the most distant black holes appear to be about a billion solar masses and formed when the universe was only few hundred million years old. That's about the time galaxies started to form. To see these links, the spacecraft will have to measure extremely faint x-ray sources at these cosmic distances.

When built and deployed, this observatory would be able to look out at a time in cosmic history called the Epoch of Reionization. This occurred a few hundred thousand years after the Big Bang and marked the emergence of light from the fog of the Cosmic Dark Ages. If possible, Lynx will also be able to detect x-rays streaming from the immediate regions of the first black holes ever to form. In addition, the spacecraft's sensitive instruments will be able to dissect the light streaming from halos of hot gas surrounding galaxies, and look at where the gas originates and what is causing it to be so hot it emits x-rays. Finally, Lynx will track down the light of very young stars, which are very bright in x-rays, and try to track down what happens to their circumstellar disks as they mature and grow.

Probing Our Origins

The topic of cosmic origins is, as we have seen, a recurring theme in astronomy and astrophysics research. The Origins Space Telescope is a mission concept that would focus on the origins and evolution, as well as the chemical abundances in planetary systems like our own. It would also study the growth of black holes and galaxies, similar to the Lynx mission, but in far-infrared light. Finally, it would study the small bodies in our own solar system and, for example, examine the role of comets in the delivery of water to various worlds. Water transport in protoplanetary disks is important, since it occurred in our own solar system and enabled the formation and support of life.

Preliminary concept work for the Origins mission has been presented, but the basic design for Origins will include a telescope sensitive to the mid-to-far infrared range, with a large mirror that will feed light to a series of cryo-cooled instruments. It would be launched to a Lagrange point position around the year 2035 (at the earliest).

It's worth noting all of these missions are highly dependent on funding and political will. The money enables the technology and pays the scientists. The political will is needed to keep scientific research as a national priority in whatever countries seek to build advanced observatories on the ground and in space. At times, it is difficult to keep the two aligned, and often science has to fight for support as every country has competing priorities for spending its tax revenues.

Connecting Past Observations with the Future

As has been mentioned elsewhere in this book, observatories now and in the future are generating incredible amounts of data. Each of the telescopes on and off the planet constantly sends back information, which gets stored on Earth. That storage, and the follow-up analysis of the data, requires an immense amount of computing capacity. For example, the Large-Scale Synoptic Survey Telescope is expected to generate more than 50 petabytes of sky data. And this is just one facility.

Not only does all this data require storage space, but also a telecommunications network to carry it all between research institutions. Even then, with extremely large data sets (such as that collected by the Event Horizon Telescope), sometimes it's faster to carry drives of information onto a jet and fly it to the analysis site.

Bigger questions loom: how do researchers doing follow-up studies of data get hold of it? Is there a centralised data repository for astronomers? Those are the challenges that led to the creation of a Virtual Observatory, which makes the data available for further study. In 2001, a decadal survey identified data-handling and sharing as a big priority for the first decade of the 21st century. Eventually, this led to the creation in the US of the National Virtual Observatory (NVO), funded by NASA and the National Science Foundation. It links together data archives for all the major NASA missions, as well as international missions, and makes it available.

For most missions, the astronomers who proposed the observations gets first crack at the data. After a certain period of time (called the 'proprietary period') it is released for public use. That's usually about a year, although for some more recent missions and ground-based observatories, some data sets are released almost immediately.

The Virtual Astronomical Observatory is a tool which puts all this data into the hands of students, educators, scientists, and others. One outcome of the effort has been the World Wide Telescope software package, originally developed by Microsoft, but now managed by the American Astronomical Society. Information from the virtual observatory also enabled Google Sky, and other packages.

Future telescope data from all the world's telescopes will eventually end up in online repositories such as NVO, allowing the world to connect to observatories in a way they haven't been able to in the past. It's not hard to imagine a future student researcher using the tools of virtual resources to make new discoveries, in much the same way that researchers have done with data and images from past observations.

The Far Future

Looking out beyond more than a decade or two in terms of technology for astronomy quickly gets us into the realm of science fiction. That's not a bad thing, because science fiction often focuses on current technology and extrapolates what could happen in the future from what we have now. Obviously, space astronomy is going to be a big thing, if all the missions that we've read about here (and more on the drawing boards around the world) are launched. The same big questions will be asked, but as the telescopes get bigger, more sensitive, and utilise new technologies, astronomers of the far future will be able to hone in on very small areas of sky over longer periods of time to capture objects and events in ever-greater detail.

So, what could astronomers be using in 25, 50, or 100 years? It's very likely that in the next half-century, the big ground-based telescopes which already exist, plus the ELT and others coming online, will still be in use. New instruments, continued maintenance on their mirrors, and advanced adaptive optics will certainly keep them relevant for decades. Of course, continued funding and interest in astronomy are also important. Taxpayers around the world have footed the bill for these scientific labs, with the expectation they will be in use as long as possible, so it's wise to maximise the investment.

Certainly there will be observatories in space. They may not always be in orbit at a safe Lagrange point or around Earth. There are far-reaching plans to use the lunar far side as an observatory reservation. The Chinese Chang'e 4 rover is a good precursor, and there will be others. Radio astronomy could benefit from such a protected preserve, as could optical astronomy. It would also make sense to figure out a way to put detectors there for high-energy research into neutrinos, cosmic rays, gamma-rays, and x-rays. Of course, such an astronomy reservation would require maintenance of the infrastructure, so it's very likely a small colony of technical staff would also be in place to keep the equipment running and up to date.

The advent of solar sails as a propulsion method offers planetary scientists another way to put instruments out in space, particularly for planetary flybys. Some suggested ideas for this technology include solar observation instruments placed in close orbit to our star, and there are some preliminary studies looking at using sails to deliver light-weight payloads to Mars and beyond. One very interesting (albeit long-range) idea is to use solar sails to place optical telescopes in space that can be linked together in a sort of giant optical interferometer. Doing so would give astronomers an 'eye' on the universe millions of kilometres across, and allow incredibly detailed views of observing targets.

As humans move out into the solar system and start exploring and living on other planets, it's logical they would also start installing telescopes on those worlds. Limited astronomy has been done from Mars, and a dedicated observatory in the Martian wilderness (with dark skies and less atmospheric aberration) makes a lot of sense. Even farther out, orbiting observatories in the outer solar system would give planetary scientists a better chance of capturing views of Kuiper Belt objects, which tend to be smaller and difficult to see from Earth (even with HST or the upcoming JWST). Of course, there are many technical challenges to confront at such great distances, and future observatories out beyond Jupiter are likely to be a very long time in coming.

Back on Earth, astronomers will continue to be the recipients of a firehose of data pouring in from space, added to the information from the missions and observatories already in place, and it's easy to see that one strand of future astronomy is going to be a data-mining revolution. There's a promising future for data analysts who can correlate and make extrapolations about celestial targets and events from the data.

A Bright, Energetic Future
Astronomy and astrophysics have come a long way in just over a century, from the days when Hale and Hubble were making cutting-edge observations using large ground-based telescopes that for their time were the equivalent of HST and JWST. Their work, done only about a century ago, was 'only' in visible light, and spectroscopy was only done with filters. The age of multi-spectral, multi-mission astronomy was likely not even a gleam in their eyes.

So much has changed and not just because astronomers are sending spacecraft out to do what was once thought impossible to manage from Earth. There's a change in the human component, too. International collaboration is the norm today, as are large teams of astronomers behind all of the big initiatives that began in the mid-20th century and continue to this day. There's enough work to do for the more than 11,000 or so astronomers known to be in the field today. They depend on technical experts, instrument builders, and people from the fields of physics, chemistry, biology and planetary science, to fully implement their grand observing plans.

The cost of doing astronomy has risen exponentially since the days of Hubble and Hale. It's not just the expense of creating the big mirrors and building the spacecraft. The telescopes can't do their work without imagers, instruments, and other infrastructure. And, as mentioned elsewhere, astronomy these days is a data-driven, byte-intensive science. The world's telescopes are collectively bringing in petabytes of information. It's getting stored, just as earlier astronomers preserved photographic plates of the cosmos for further study. There's a whole discipline of data-driven research, where astronomers comb through observational records to learn about earlier observations, correlate their information with older data, and perform extensive analysis.

The infrastructure of modern astronomy is expensive. That's why multi-national collaborations are forming – to spread the cost. The upside is these teams also bring in experts from around the world to concentrate on the big questions everyone wants to answer. Finding money for new observatories is a challenge to every country that wants to build them. In a time when budgets are constrained, scientific decisions have to be completely justified to the taxpayers of the countries involved. In the long run, however, basic research has always paid off in technological advancement that benefits not just scientists, but entire societies. The money spent on observatories (whether in space or on Earth) doesn't get lost in space or out in a desert. It goes to pay salaries of workers, who in turn spend it on homes, clothing, food, and the other necessities of modern life (including taxes!).

Astronomy has also long been called a 'gateway science' to other sciences and technical advances. As we've seen in the history of astronomy, stargazing led to advances in timekeeping and calendar-making. Eventually, the need for a better 'view' of the sky led to the development of measuring instruments and telescopes. The telescope itself required improvements in optics and materials science.

When the science of spectroscopy was developed, it brought chemistry into the mix and allowed astronomers to identify the chemical elements in stars. The study of other worlds (starting with the Moon and Mars), brought geology and geographical landforms to astronomy and created the discipline of planetary science. In recent years, the search for life and the biological markers of life have involved the life sciences community, and led to the discipline of astrobiology. As we've seen with the current crop of neutrino and cosmic ray detectors, the study of these fast-moving energetic particles from space have united astronomy with particle physics. And, the study of stars created the science of astrophysics. So, from being humanity's first science, astronomy can take credit for a lot more than the simple study of stars. It's now a study of the universe and all the things that make it up.

The creation of newer, bigger, faster, further looking instruments is part of astronomy's heritage. Looking to the future of astronomy is, in a very large way, looking at the future of humanity's outreach to the stars. The Moon and planets are within our reach for in-situ exploration, with Mars and chosen asteroids being marked as targets for crewed missions. Travelling to the nearest stars is a bigger challenge. We don't travel at the speed of light – nowhere near it. So, any star-exploration missions by humans will be multi-generational ones. Even the nearest star – Proxima Centauri – is decades away at best. While we may not be able to travel to it soon, we can study stars from Earth with our ever-improving observatory assets. That's one of the biggest gifts astronomers have given us: an up-close look at the universe from the relative comfort of our home planet.

8

OBSERVATORIES IN BACKYARDS, EDUCATION, AND POPULAR CULTURE

Throughout this book, the focus has been on observatories as those places that, throughout the ages, have been dedicated to researching the large questions: what's out there? What is the nature of the planets, stars, and galaxies? Are we alone in the universe? How did life arise? How did the universe begin? How will it end? What else is out there? For members of the public, observatories have may seem like 'temples to astronomy research' and places where (literally) 'out-of-this-world' discoveries are made. Lowell Observatory, for example, brought knowledge of Mars and Pluto to turn-of-the-19th-century audiences. Its pair of discoveries spurred new generations of stargazers and scientists to turn their attention outwards. Even before that, stargazers who – by today's standards would be classified as 'amateurs' – were making momentous discoveries. People such as Galileo (who was by training and interest primarily a physicist and engineer who also loved looking at the sky), or William Herschel (who was trained as a musician, but fell in love with astronomy) and his sister Caroline Herschel (who also was trained in music and dressmaking, but like her brother also was fascinated by astronomy), may not have been the equivalent of today's professionals. However, their dedication to sky-gazing got us to where we are today: exploring the universe with all manner of instruments. Many a professional astronomer started out with a backyard-type telescope not much bigger than Galileo's, gazing up at the planets and stars in wonder and admiration. If they were lucky, they had access to an observatory while still at school and later on, in college or university; maybe they visited a university facility during an open house, or made field trips to a public observatory and science centre. So, in addition to being devoted to research, observatories are also cultivators of young minds and fertile fields where the seeds of research start to grow.

The Backyard Observatory

Amateur astronomy is, by long (but possibly changing) definition, astronomy done by those who love stargazing and aren't necessarily making a living as a professional astronomer. It's a rather simple definition, but if we take it seriously, then much of astronomical history has been done by amateurs. The difference between 'professional' and 'amateur' observer is a distinction that is changing as more people can get their hands on good quality telescopes for their interests. Not only do individuals own and operate their own observatories, but many astronomy clubs and associations do, too. Many who 'do' backyard astronomy also contribute to science, continuing the tradition that began with Galileo.

These days, the line between 'pro' and 'am' astronomy is an exceedingly fine one. Since at least the last part of the 20th century, dedicated amateur observatories have participated in scientific pursuits in coordination with professional researchers, using telescopes amateurs of previous decades (and pre-20th-century professionals) could only dream of having. And, many more have been indulging themselves in the joy of searching out planets, stars clusters, nebulae, and galaxies simply for their own enjoyment and self-education.

Most amateurs do their observing in visible light. Some also conduct radio astronomy experiments, using easily built kits. A few amateurs do simple spectroscopy in their backyards, although to go much further and make spectrographs is a far more complex task. Many also do astrophotography, and with the advent of the affordable astronomy CCD cameras, they can turn out some very beautiful images which are regularly published in magazines, while a few of the best sell prints of their images to collectors. The easy availability of good-quality CCD cameras and affordable telescopes has sparked a boom in imaging. Users of social media and the web are very likely familiar with friends and acquaintances who indulge this hobby and post their work for all to see.

Today, there are amateurs who have research-worthy telescopes installed in professionally built domes in their backyards or on other private property. It turns out observatory domes are increasingly being put into newly built homes, but even those who can't afford a custom-designed home have found ways to indulge in their interests. People who don't have access to dark-sky sites can, as well, get access to research-grade telescopes to do stargazing and photography for a small fee.

Amateur astronomy is a world-wide phenomenon, with groups of like-minded telescope users and builders in nearly every country. In the United States, amateur astronomy really took off after the Second World War. Before that time, telescopes were quite expensive and usually only found in the homes and private observatories of relatively wealthy people. Yet, people were fascinated by celestial events such as eclipses and the appearance of comets. There was a huge interest in the sky.

In the 1920s, people such as Russell Porter (who worked on Mount Wilson's Hale Telescope) spent his time designing and building telescopes and teaching people how to build their own. Porter is often referred to as the 'father of amateur astronomy' in the US and shared his knowledge freely via articles in *Scientific American* and other publications.

After the war, however, people began building their own telescopes, and demand arose for mass-market instruments. The war had spurred a need for optical instruments, and afterwards, highly trained optical experts turned their skills towards making telescopes. In addition, in the 1950s, smaller mass-produced telescopes were being made in Japan and sold in the US, spurring an interest in the skies. By at least one estimate, hundreds of thousands of Americans got involved in amateur astronomy by the early 1960s and the dawn of the Space Age. Today there are astronomy clubs across the country, some with private observatories. Hundreds of thousands of telescope owners regularly go out to observe, some of them quite skilled and interested in furthering their own knowledge of the sky.

The US is not the only country with a mass-market interest in astronomy. In Japan, for example, interest in the sky and astronomy dates back several hundred years. Modern interest in the sky and the growing interest in observing by Japanese amateurs really dates back to about 1865, when the country's professional astronomers took a strong interest in western astronomy. In the 20th century, amateur-professional cooperation seems to have begun rather early on. In 1908, the Nihon Tenmon Gakki group (The Astronomical Society of Japan) was formed with an eye on informing the public about astronomical discoveries. Certainly the appearance of Comet Halley in 1910 further spurred interest in the sky, just as it did around the world.

The advent of the Second World War had an adverse effect on Japan (and of course was disruptive to the US and Europe), so it wasn't until the post-war years that Japanese amateurs (and professionals) could return to their interest in the sky. This has continued since those years. Japan has dedicated itself to science education, and has built astronomical learning centres which combine observatories, planetariums, and science centres where members of the public can learn about the stars and planets. People throughout Japan flock to dark sky sites (which are rare in that country) and to other countries to indulge in observing and astrophotography. Many of the world's best sky imagers live and work in Japan, perfecting their techniques even as they learn more about the sky.

Amateur Equipment

So, what kinds of equipment are amateurs using these days? Whole books exist that examine that question in great detail. Obviously, they're using telescopes in a range of prices and designs. Some users prefer refractors, others use reflectors. All need to be mounted on sturdy foundations (usually what's called 'pier mount'), and protected from the elements and other intrusions. What's most important are good optics, computer controls (to get fancy), some kind of image-capture device, and a dark sky site from which to observe and work.

A quick look at the ads in an astronomy-related magazine such as *Sky & Telescope, Astronomy, Popular Astronomy, Astronomy Now, L'Astronomie* and others, show a plethora of different instruments.

Companies such as Astrophysics, Celestron, Meade, Obsession, Questar, Takahashi, among many others cater to the small-telescope market. But 'small' varies in size; most amateur telescopes' apertures range from a few millimetres to Dobsonian 'light buckets' 1.7 metres (70 inches) across. (Dobson's contribution to astronomy is dealt with below.)

Housing the Telescopes

Backyard observatories can be as simple as a telescope on a deck or a picnic table or even a large empty grassy spot. What's important is that it has a relatively unobstructed view of the sky, away from direct sources of nearby light. Just as early telescope builders found, however, today's serious observers often want to leave their telescopes set up, protected from the elements (and other incursions).

Amateur observatories can be housed in something as traditional as a dome (and they come in various sizes and for varying budgets) to shed-like buildings with roll-off roofs. Most amateur installations have electricity, lighting (low-light), and computers with network access to either their home LAN or the Internet. Some have warming rooms, refrigerators, and other amenities to help make those all-night sessions more comfortable. In the past, before the advent of digital imaging, the really advanced amateur observatories even had darkrooms where the user could develop their own film from a night's astrophotography.

Making and Using Telescopes

Some amateurs build their own telescopes, particularly if they want an instrument of a certain size and are interested in the challenge of building a unique one. There are many different ways to approach telescope making, which we won't get into here, but it does start with mirror grinding, which is something almost anyone can learn to do. There are clubs that teach the methods, as well as individuals, such as the late John Dobson. He developed the Dobsonian telescope, with a Newtonian reflector he wanted anyone to be able to build

John Dobson was an amateur astronomer for many years. He spent more than two decades of his life in a monastery. There, he became interested in the universe, and turned his attention to telescopes.

After leaving the cloister, he founded the San Francisco Sidewalk Astronomers to foster public interest in stargazing. In the 1960s, he began working on his telescope design, and touted it at star parties and public stargazing events. He could often be found showing young people how to grind a mirror and advising more advanced telescope makers on the finer points of these self-built telescopes. Today, the Dobsonian and variants on his original design are found in many a backyard observatory, and at star parties, alongside high-end refractors and affordable Newtonians purchased ready-made.

Among the club-oriented observatories is the Amateur Telescope Makers of Boston (Massachusetts) organisation. Located about 48 km outside of Boston, this remote site, situated next to MIT Haystack Observatory, is home to telescope makers, astrophotographers, and observers. It was founded in 1934, with advice from Dr Harlow Shapley of Harvard College Observatory. It's one of the oldest and largest such organisations in the US. Not only do the ATMOBs (as they call themselves) hold monthly meetings, the clubhouse is open each Thursday and Saturday for members to work on their own telescopes or use the club's equipment. The Saturday events are open houses. The mission of the observatory is to help people learn astronomy, build or use their own telescopes, and construct their own observatories (if interested).

Another club where telescope making and mirror-grinding are taught is the Stellafane Amateur Telescope Makers group in Vermont. They also host an annual star party and, for three days, it draws observers, mirror grinders, and lecturers from around the world.

A World of Amateur Observatories

It's an unending task to tally the world's amateur observatories. Various places that keep track of them on the web list hundreds with active websites. Not everyone who has a backyard observatory, however, publicises that fact. Plus, not all are individually owned; some belong to clubs and other organisations, too. For security reasons, many amateurs do not want to give too much information out about their sites, but others do.

For observers who want to live the astronomy lifestyle, purpose-built communities are springing up in such places as Arizona, which bills itself as the astrotourism capital of the United States. They exist near Flagstaff, which was the first Dark Sky City in the world. The southern part of the state is home to Arizona Sky Village, an astronomy community featuring places where landowners can build homes and observatories under some of the darkest skies available.

Inside Look at a Personal Observatory

Often a personal observatory reflects a wide variety of the owner's interests. Richard S. Wright Jr, a software developer, writer and astrophotographer, operates a personal private observatory with some friends/partners at a location in Okeechobee County Florida called Stardust Ranch. His love for astronomy began very early, practically before he could even read or write. Today, Wright's interests have grown to encompass telescope building, coding, astrophotography and much more. He has a collection of telescopes and mounts that he has used for astrophotography as well as his regular 'job' of writing software for telescope controllers and planetarium programmes. His software helps run observatories and amateur telescopes around the world, including a lot of camera plug-ins and device interfaces. Such a dream job gives him access to a wide variety of hardware most amateurs could only dream of having.

As part of his work, Wright has used just about every design of telescope anyone could imagine for both visual astronomy and astrophotography. After all this experience, he prefers simpler instruments and setups, the 'less is more' approach to telescope design. He loves refractors best, and for him, no telescope design is better than a high-quality premium refractor when it comes to contrast, and clean tight stars and object details.

Since Wright also does a great deal of imaging, his favourite imaging scope is a 15-cm (5.9-inch) Esprit refractor. For pure ease of use, he often turns to his biggest telescope, a 30-cm (11.8-inch) Newtonian. For visual observing, he uses a 35-cm (13.7-inch) telescope, and a smaller 17-cm (6.6-inch) instrument, which is very portable and easy to use on trips. Wright also has a small telescope equipped for solar observing. He is typical of many avid observers who have several different telescopes depending on what they want to observe and/or photograph.

Observing and Learning Together

Astronomy clubs and societies promote stargazing among their own members and reach out to the public. These organisations exist around the world, in nearly every country and unite stargazers by way of meetings, lectures, and star parties. For example, The Royal Astronomical Society of Canada is a country-wide organisation that began in 1868 as a gathering of amateur astronomers. Today, it has more than 5,000 members spread across twenty-eight centres where interested skygazers can learn the rudiments of amateur astronomy, build and use telescopes, and share their passion with the public. In England, the British Astronomical Association serves a similar role, with members across the UK (and around the world) who observe and share their knowledge with others. The organisation offers 'basics' workshops as well as information for advanced amateurs, and is organised into observing sections focused on specific types of targets.

Sky Access for All

Not everyone can build an observatory in their backyard (or other private property), but still may want access to a good observatory. The phenomenon of private for-profit observatories for rent has arisen in the last two decades. These internet-accessible facilities offer access to the sky for long- and short-term projects to anyone who wants to use them.

One of the more interesting and ambitious amateur project, called the Virtual Telescope project, was started by Italian astronomer Gianluca Masi long before he ever went to graduate school. He first fell in love with the sky as a young person in the 1970s. He got his first telescope in 1983, a small 6-cm (2.3-inch) refractor. It allowed him to look at the planets, the Moon, and the Orion Nebula. Two years later, he got another 150-mm (6-inch) reflector with a computerised mount, and began his search for deep-sky objects. At about the same time, he took up astrophotography, which was becoming more accessible to amateur astronomers.

In 1997, Masi founded the Bellatrix Astronomical Observatory in Ceccano, Italy, about 2 hours outside Rome. He studied astronomy at the University of Rome, earning a Ph.D. After extensive observing experience, Masi started the Virtual Telescope project in 2006. It provides online access to two telescopes for interested observers. They are a Celestron C14 and a PlaneWave 43.1-cm (17-inch) telescope. They can be outfitted for observation and astrophotography, and are available to both amateur and professional use. The observatory also offers public online observing sessions for specific events such as eclipses, comet-chasing, and other celestial happenings. Most services are free to access, but some activities do require fees to help keep the observatory going.

The rise of Indian space exploration is coinciding with a wave of interest in amateur astronomy in that country. There are a number of clubs and societies, along with planetariums and observatories open to the public. The Stargate Project is a private venture which offers access to observatories for students and tourists; it teaches the rudiments of stargazing, astrophotography and astronomy. They also conduct astronomy adventure tours. One project brings students and other visitors to do observing at a dark-sky site in a national park, where they can also do astrophotography under the guidance of an expert.

Learn to Explore Space

Most people know where they were on 11 September (9/11) 2001. On that day in the United States, hijacked planes crashed into towers in New York, the Pentagon in Washington DC, and a farm field in Pennsylvania. It was a horrifying time. A young man in New York City watched as one of the towers fell, killing his college friend, Blake Wallens. And, on that day, Michael Paolucci, who had grown up reading the works of astronomer Carl Sagan, decided to use his inspiration to create a memorial to his friend. The memorial is Slooh.com, a remotely operated robotic set of telescopes that have spread out from the original base at Tenerife in the Canary Islands to telescopes at La Dehesa in Chile. This gives users access to both northern and southern skies. In 2020, there will be telescopes available on line from three continents, 24 hours a day, seven days a week.

Slooh's motto is 'Learn to explore space' and every telescope works through a cycle of 'missions' each night. Every 5–10 minutes, the telescopes in the network focus on different objects as part of a mission. These targets are selected by the organisation's resident astronomer, or through reservation requests by members who pay between $20 and $100 to join as individuals. There are separate levels for club and classroom users.

Slooh operates through an easy-to-use web-based interface and functions as a citizen observatory. Since its public launch in 2003, users have connected to the telescopes to study near-Earth asteroids (NEAs), look at and image stellar outbursts, discover supernovae, novae, comets, and distant asteroids. Slooh has also been used in professional-amateur collaborations for comet studies, and large-scale observation efforts. The NEA tracking programme has been wildly successful, and according to the network website, members have submitted nearly 10,000 measurements of asteroids to the International Astronomical Union (IAU)'s Minor Planet Center at the Smithsonian Astronomical Observatory in Cambridge, Massachusetts. Slooh has partnered with NASA on asteroid studies, and was awarded a grant by the National Science Foundation to support its efforts in astronomy observations and education.

Many Eyes Make for a Bigger View

The Los Cumbres Observatory (LCO) is another network with facilities around the world. It is a non-profit observatory with access for the public and professional observers, and maintains multiple facilities at Cerro Tololo, the South African Astronomical Observatory, Siding Spring in Australia, McDonald Observatory in Texas, Haleakala Observatory on Maui in Hawaii, and Teide in the Canary Islands. There will be new facilities added in Tibet and Israel. The network consists of two 2-metre (6.5-foot) telescopes, nine 1-metre (3.2-foot) telescopes, and seven 40-cm (15-inch) instruments. All the telescopes operate together as an integrated unit, tied together by computers and scheduling software, and perform observations on demand from the supporting users. It can perform what is called 'time domain observing' which allows the telescopes to watch continuously as events such as a supernova explosion, take place.

LCO is supported by grants from various foundations and companies, as well as individual donors and users who buy network time for observations. The US National Science Foundation made an award that supports US astronomers for a certain amount of time each year. The network is also open to educational users, and supports a number of educational outreach projects around the world. Its affiliated group, called the Global Sky Partners, is an umbrella organisation coordinating more than a dozen special programmes. Their educational initiatives include a project called '100 Hours for 100 Schools', which supports student observing projects. The LCO team also is working on bringing astronomy to underrepresented groups of students through school projects, after-school programmes, and other events.

Astronomers from institutions around the world have used LCO to search for exoplanets, study asteroid families, search for potentially hazardous asteroids, look for supernovae, look at transient events near the centers of distant galaxies, look at cataclysmic variable stars, and much more.

Los Cumbres is headquartered in Goleta, California. It is the brainchild of engineer Wayne Rosing, who set out to build out a complete global network of telescopes. Its data and images are used by the requesting observers first, and then are made publicly available.

Experiencing the 'Group Observatory': Star Parties
Every year, amateur astronomers come together at star parties. In a sense, they put together a self-assembling astronomy village, a temporary observatory consisting of dozens or (in the larger star parties) hundreds of different kinds of telescopes. Star parties happen all over the planet, usually during the summer season, but can occur at other times. They feature stargazing opportunities, guest speakers, and sometimes even a vendor area where people can try out eyepieces, look at newer model telescopes, test software, and buy other astronomy-related gear. It's worth mentioning a sampling of star parties here, only a few of the many that occur.

In the United States and Canada, there's generally at least one star party a month, often more. The Winter Star Party, held in the Florida Keys each February, is a limited-attendance event (due to the fact that the island it's held on can support a relatively small number of people). The Texas Star Party is held each spring at a dark sky site near Fort Davis, and regularly draws people from around the world. Stellafane in Vermont, is probably one of the oldest gatherings; each year in August, it draws telescope makers and stargazers for a long weekend of talks, mirror grinding, and other activities.

Starfest, held each August, is one of Canada's largest gatherings of amateur observers. Like many other star parties, it offers camping and dark skies, located only a couple hours away from Toronto. There are observing sessions, guided sky tours, lectures and informal talks by professional astronomers from around the continent, telescope building and software workshops, and children's activities. Like many of the larger star parties, Starfest also encourages commercial exhibits from astronomy-related vendors.

In Japan, the National Observatory of Japan holds regular star parties twice a month at its facility in Mitaka. The organisation makes a 50-cm (19-inch) telescope available for public use, as well as a collection of smaller instruments. Participants usually get to observe the Moon (if visible), planets, stars, and some deep-sky objects.

British observers are no strangers to star parties. There are around 15 events each year that draw amateurs from around the UK. For example, the Kielder Star Camp brings together amateurs twice a year in one of the country's darkest sky sites. People treat it as a camping event as much as a stargazing opportunity, and, weather permitting, up to 200 attendees can converge on the area for each event. There are usually some lectures presented at nearby Kielder Castle that include everything from personal observing experiences and equipment demonstrations to the latest astronomy discoveries and space missions. The camps are organised by the Sunderland Astronomical Society, a group of amateurs who banded together 1993 to stargaze together. The Birmingham Astronomical Society organises its annual Winterfest Astro star party, usually in November. It's held at Kelling Heath in Norfolk and, like Kielder, is a camping event. The society itself was founded in 1950 as a club for both professional and amateur observers. It maintains a small observatory at Aston University, and offers mirror-grinding instruction, as well as access to the online Astronomy Society Academy courses conducted by educator Alastair Leith. The Winterfest party is one of its main outreach events each year.

Australia is home to a number of dark-sky sites free of light pollution. To take advantage of the southern skies and the pristine viewing conditions, the Astronomical Society of New South Wales holds the annual South Pacific Star Party each year. The observing site belongs to the society, and is located a few hours' drive outside Sydney. It usually brings together up to 400 observers and their telescopes for several days and nights of astronomy. There are daytime workshops and lectures designed to help telescope makers advance their craft.

Professional-Amateur Cooperation and Citizen Science

With the rise of star parties, the continuing spread of amateur astronomy and the rise of good-quality telescopes in the hands of motivated amateur observers with backyard observatories, it was only a matter of time before professionals and amateurs would get together to 'do good science'. The story of this came about is an interesting one. For many years, amateurs were 'looked down upon' by their more-educated peers in academia as not being sufficiently well trained (as one formerly snobby professional once confessed to the author). There have been notable exceptions over the years. For example, in Australia, professional and amateur observers worked together to educate the public about the things they were seeing through their telescopes. But, that was largely an educational outreach project, and not necessarily a joint science venture. A better example of close cooperation between amateurs and professionals exists with the American Association of Variable Star Observers (AAVSO). This is an international group founded in 1911, which has been coordinating observations of variable stars with other observers and observatories (including space-based ones), ever since. It's open to anyone who can help with the ongoing tracking of variable stars, and bringing those observations to astronomers who either don't have the time or access to telescopes to do the constant studies required to capture the variations in brightness of stars, including our own Sun. AAVSO's Solar Section is dedicated mainly to monitoring sunspots and is well-suited to users with smaller telescopes (and solar filters!).

Why Variables Are Important

Variable stars change their brightness and this can vary over short periods of time or take days or weeks to occur. Their variability might be only a very tiny fraction of a percent, or so noticeable that a naked-eye observer can see it. Stars vary in brightness for a number of reasons. Perhaps a star dims a bit because another star passes in front of it. Or, it might vary in its brightness because it is swelling or shrinking (a so-called 'pulsating variable'). It could be a member of a binary pair of stars orbiting a common center of gravity. As they orbit, one star could 'steal' mass from the other, causing it to brighten and the other star to dim. In recent years, astronomers have noticed tiny dimmings that indicate the presence of a planet orbiting a star.

To understand these stars and why they are variable requires extensive observation time. Professional astronomers interested in them cannot always get access to telescopes to make constant observations of these stars, but amateurs are uniquely suited to do this. So, AAVSO began recruiting observers to make long-term studies of these stars. They use a variety of methods to do this, all directed by the organisation. They can do visual observing, take photographs using CCD cameras, and make other measurements that help tell professional astronomers about the behaviour of a variable. All the observations are stored in a database, and are used to schedule observations (both on Earth and from space) of certain stars, to compare data from satellite and ground-based observations, to help analyse a given star's behaviour, and to fill in theoretical models of variable star activity. The AAVSO's work is an excellent example of professional-amateur cooperation in astronomy research.

Expanding Pro-Am Cooperation

In the later years of the 20th century, attitudes began to change, in part due to the data provided by such groups as AAVSO. Astronomers began looking at their amateur cousins with new interest. One good example comes from the Comet Halley Watch, created by NASA and spearheaded by astronomers who realised that there were acres of good glass out there being used by amateurs which could be put to use to look at the comet. More to the point was the fact was that time on bigger, research-grade telescopes, was at a premium (and still is, actually). So, it was tough to get time to look at something like a comet. As a result, a team of scientists put together the Watch and sent out requests for participation to amateurs around the world. The response was very good, with thousands of images of the comet throughout its apparition pouring in for analysis. (Note: the author was one of those doing the research.)

The success of the Watch led other scientists to seek out amateur partners in what's been called 'Pro-Am Cooperation'. In many cases, the amateurs supply the observatories and imaging services. The professionals supply guidance and requirements for the specific objects they want to study. Resulting papers are supposed to credit the amateur as a contributing author or partner.

In the early 1990s, the author and a team at the University of Colorado created the (now defunct) Ulysses Comet Watch Network. It was a follow-up to the Halley Watch network, using many of the same observers. With the advent of digital imaging, however, instead of getting negatives, prints and slides, the research team received TIFs and lightly-compressed JPGs, as well as other types of digital images.

Here's how the process worked: a comet would be discovered that fitted into the parameters of the team's interest. For example, it had a perihelion reasonably close to the Sun, it had an orbit that 'sampled' a wide range of latitudes of the solar wind, and it would be in those latitudes at the same time as the Ulysses spacecraft. The mission was in a roughly polar orbit around the Sun, so the idea was to look at a comet when it was 'seeing' or experiencing the same part of the solar wind where the spacecraft was traveling. The team would send out observing notices to participating members and they would observe the comet and dutifully return images with specific dates, times, and other location information. With both the Ulysses Comet Watch and the Halley Watch, the effort benefitted from both professional and amateur observers, as well as individual and group efforts. One of the most prolific of the comet imaging teams was the Volkssternwarte Frankfurt, which had observers and astrophotographers from its own comet section submitting images.

There have been other comet watches over the years since then, including the Professional-Amateur Collaborative Astronomy (PACA) project, funded by NASA. Over the past few years, the leader of the team, Dr Padma Yanamandra-Fisher, has gathered a cadre of experienced comet observers from around the world. One of its targets was Comet 67P/Churyumov-Gerasimenko, which was studied by the Rosetta mission. Amateurs using home observatories and traveling telescopes imaged the comet and shared the images as part of a larger outreach effort by the Rosetta team. Other PACA groups have focused on outbursting comets, as well as small worlds as adjuncts to mission-based scientists who are studying the same objects.

Backyard Astrophysics and Other Pro-Am Projects

Comets aren't the only objects that attracted pro-am cooperation. In recent decades, amateur observers have turned their telescopes to planets, nebulae, clusters, and galaxies, in addition to variable stars. In addition, amateurs with good equipment have been studying microlensing events, blazars, and other oddities in the cosmic zoo.

The Center for Backyard Astrophysics (CBA) is a project which unites astronomers at all levels around the world. Founded in the 1970s as the Center for Basement Astrophysics, the organisation has grown and renamed itself in honour of observers who don't actually work in basements, but observe from their backyards. The central interest for CBA is cataclysmic variables, often referred to as CVs for short. They are stars which erupt periodically and unpredictably. CVs can be novae, symbiotic stars (two stars in a symbiotic relationship), and recurrent outbursting stars. These objects are usually binary stars orbiting very closely around a common center of gravity. Material from one partner spirals onto the surface of the other, and makes an accretion disk, which is usually very hot and quite brilliant. Occasionally, the system bursts out in brightness, and that's part of what astronomers hope to understand about these objects.

It's important to track the brightness changes in cataclysmic variables and for the purpose, well-equipped amateur astronomers are a good choice. The CBA requires observers to have a good telescope, usually at least an 8-inch aperture, a computer, a detector (imaging device), proper software, and star charts. The group sends out periodic observing campaign notifications and the members get to work doing their observations. The outcome of all this work continues to be a better understanding of the orbits of close binaries, the formation and activities in their accretion disks, and the mechanics of brightenings and outbursts.

Blazars, Quasars, and Lensing, Oh My!

The late Canadian astronomer Paul Boltwood became well known as one of the first amateurs to do long-term studies of blazars. These are the active cores of galaxies, known as active galactic nuclei (AGN), with jets blazing away from the center. His observations from his Ottawa-based observatory using a 40-cm (16-inch) telescope and a camera he designed himself aided professional astronomers who were studying these objects.

Not only are AGNs and blazars bright enough to observe through amateur telescopes, so are quasars. These objects are extremely far away, but incredibly bright across most of the electromagnetic spectrum. With today's modern telescopes and cameras, quasars are now within reach of amateur observatories. The Astronomical League, which has member societies across the United States, has a specialised observing programme for Active Galactic Nuclei, including quasars. Participating members are given specific requirements for observations, which they then turn in for professionals to use.

Things That Go Flash in the Night

Meteors have long fascinated skygazers, and there's a professional-amateur collaboration bringing together meteor watchers and researchers. The American Meteor Society was founded in 1911 and encourages people to observe, monitor, collect information, plot, and report on meteors and fireballs. Advances in cameras, telescopes, and other technology have enabled many observers to capture meteors 'in flight', as well as to monitor known meteor showers that occur each month. It has four different programmes that members can participate in: visual observing, video recording, radio observing (tracking them by radio signals reflected off of a meteor trail in the upper atmosphere), and meteor spectroscopy (studying their light through a spectroscope).

Citizen Science

While not necessarily depending on an observer having a telescope, citizen science has, nonetheless, become a huge factor in many science research projects. These are generally projects where data has already been gathered, but help is needed from many observers to spot such things as features on Mars, or exoplanets, or gravitational lenses, for example. The most extensive site for such projects is Zooniverse. org, which has its own astronomy section. Some projects at major observatories or space missions have data sets there, including a muon hunters section from VERITAS, a space warps search project supported by the Subaru Telescope using data from its Suprime-Cam imager, and a planet hunters' project using data from the TESS spacecraft currently in orbit. These projects take the time to train their observers, and trained amateur astronomers are a plus.

Think of citizen science as an adjunct or even an extension of an observatory's work. Some analyse the data for entirely novel purposes. For example, the Content-based Object Summarization to Monitor Infrequent Change (COSMIC) project uses images taken of Mars land features to enable machine learning algorithms that could be used aboard future spacecraft. By helping with this training, users are teaching the algorithm to search out scientifically important landforms on the Red Planet. Then, when it's installed on a spacecraft, it will know whether a given landform matches something in its database that was flagged as 'a landform connected to impacts' or 'this canyon looks like a flow feature where water once existed'.

Another project on the website is called Galaxy Zoo, and is one of the oldest of the citizen projects listed there. It's been in operation for over a decade and asks users to sort galaxies by their shapes. People may also be asked to note if a galaxy seems to be interacting with another one. Galaxy classification is an important part of learning how these stellar cities form and evolve. Images have come from the Sloan Digital Sky Survey in the past. Currently, the Zoo is using images from the Dark Energy Camera Legacy Survey, conducted from Cerro Tololo in Chile.

In 2019, astronomers announced a major new study based on galaxy classifications performed by Galaxy Zoo volunteers. Based on work done by users on more than 6,000 galaxies, scientists discovered there is no correlation between the extent of a spiral galaxy's central bulge and how tightly its arms are wound. The ground-breaking work overturned a long-held idea first proposed by Edwin P. Hubble, who suggested, based on his own much smaller survey, that spirals with large central regions (bulges) had very tightly wound spiral arms. This kind of crowd-sourced analysis is a major feature of citizen science based on the mountains of data which observatories are providing each year.

Telescopes in Education

The next generations of astronomers are already in our schools. Some are just starting their path to learning the sky. Others are in college programmes or graduate school, getting hands-on experience with the tools of the trade. In many countries, astronomy is part of the science curriculum, along with other science, technology, engineering and maths (STEM) studies. In the United States, for example, astronomy topics are woven into the various curriculum guidelines and it's left up to schools and educators how to present and teach the materials. In many places, trips to science centres and planetarium facilities supplement classroom lessons. Not every school has a telescope, although there are a number of school observatories out there.

The Clay Center for Science and Technology at Dexter Southfield School in Brookline, Massachusetts, in the US, is a good example of a school with a dedicated facility for astronomy. It has an observatory which allows students at the school to conduct scientific observations of planets, the Sun and Moon, and stars. Its programmes are not limited to students; the center also provides outreach to local groups such as Scouts, science clubs, and other organisations that are interested in science education. There also open scope nights each week, allowing visitors from around the region access to the center's professional-grade 60-cm (23-inch) computerised reflecting telescope, plus a 17-cm (6.6-inch) refractor and camera. In addition, in 2011, the center installed a planetarium for its classes.

The Wilton Community School district in Iowa put the finishing touches to an observatory for its students in 2019. Called the Wilton Observatory, it offers students a hands-on opportunity to explore astronomy with a telescope they can learn to run. The project, which cost about $90,000 in donations, allowed the observatory's founder, Grant Harkness, to get an automated Meade telescope. The observatory has software packages students can use as they study stars, planets, comets, the Sun, and other celestial objects. Use of the observatory is being written into every grade level so that children in the area can learn the basics of astronomy from an early age.

There are many such facilities, and they're aren't limited to the United States. The National Schools Observatory in the UK, administered by the Liverpool John Moores University offers free access to students via their robotic instrument, the Liverpool Telescope. The idea is to bring the universe to local school pupils and their teachers. Up to 10 per cent of the telescope's observing time is allotted to schools in the UK and Ireland. It enables students to learn how to use the telescope and do real science with their assigned time. It also provides a sort of internship programme for teenagers, and also conducts teacher training sessions.

Astrobazas to Space

One of the most ambitious country-wide programmes for astronomy has taken shape in Poland. With that country's interest in space through the Polish Space Agency, which is part of the European Space Agency, the stage was set to build a network of fully equipped observatories for students from the ages of seven to twenty-one. The observatories are called Astrobazas and each building is near a local school. In many places, the parents helped build the facilities, and the students maintain and use them. Each Astrobaza features a 35-cm (13.7-inch) Meade telescope as the centrepiece. Other pieces of equipment include a Coronado PST 40/400 solar telescope, 15x magnification binoculars and CCD cameras optimised for astrophotography. The instruments allow students to observe everything from the solar system out to distant galaxies. The project also provides state-of-the-art equipment for astronomy and weather observations, computer coding, and other related activities. The activities are woven into school curricula and satisfy the country's Math, Science, and Technology (MST) standards.

University and College Observatories

There are literally hundreds of college and university observatories around the world. They serve as the training grounds for undergraduate and graduate students interested in pursuing a career in astronomy, astrophysics, planetary science, and other related fields. The oldest known one is the Leiden Observatory, founded in 1633.

In addition to their own in-house equipment, many astronomy and physics departments at colleges and universities participate in collaborations with other schools and institutions to gain access to the latest advances in observatories at Kitt Peak, Mauna Kea, Chile, Australia, the Canary Islands, and other locations. Some university facilities, such as Harvard College Observatory, Yerkes, Lick, and others already mentioned earlier in the book, have served generations of students seeking careers in astronomy since the 1800s. Others, such as the Palomar Observatory, twin Kecks in Hawaii, and the University of Colorado's (CU) Sommers-Bausch Observatory are typical of the type of facilities built in the 20th century to serve growing astronomy research activities.

Sommers-Bausch Observatory (SBO) is part of CU's Department of Astrophysical and Planetary Sciences on campus, and provides hands-on training for both undergraduate and graduate students in the department. Its main instrument is a 60-cm (24-inch) Boller & Chivens telescope and for many years it was the only one under the dome. Students have used it for astro-imaging, spectroscopy, and photometry. The first instrument installed at SBO was a 26-cm (10-inch) Bausch refracting telescope, put in place when the observatory opened in 1953. It has since been retired. The observatory has several other telescopes, plus a heliostat and imaging equipment. Sommers-Bausch hosts the university's astronomy club, plus local astronomy and space clubs. It periodically holds open-house viewing sessions for the public and offers tours.

Radio Astronomy Education

Haystack Observatory in Westford, Massachusetts, is another university facility dedicated mostly to radio astronomy and related subjects. It is owned and operated by the Massachusetts Institute of Technology. While this facility is heavily involved in world-wide efforts toward radio arrays, it also conducts ionospheric radar research, and offers educational opportunities for both graduate and undergraduate students. At the undergraduate level, Haystack involves students in ongoing research projects in radio astronomy. One interesting project developed at the observatory for student and teacher use was the small radio telescope (SRT) kit. It gives students and their teachers a hands-on introduction to radio astronomical observing techniques. Using the kit, they can do observations of atomic hydrogen and solar emissions. There are many SRTs installed on college campuses around the world, and it's bringing the techniques and science of interferometry to the undergraduate classroom.

Haystack has a 37-metre (121-foot) radio telescope available to students across the US for research use. It's remotely accessible and gives these observers a chance to learn more about doing radio astronomy research. Observations using the telescope are written into undergraduate curricula at all levels, ranging from demonstration experiments in entry-level astronomy courses to projects at higher levels. In the past, students have used the telescope to study emission from a variety of molecules including ammonia, methanol and silicon monoxide.

In the British Isles, universities and colleges such as University College London, Cambridge University, and the Royal Observatory Edinburgh (among others) have been teaching and training astronomers using their own observatories as well as facilities around the world with which they are affiliated.

The UCL Observatory (UCLO), located at Mill Hill in London, is a teaching facility run by the department of Physics and Astronomy at University College London. It was conceived of and built under the aegis of Astronomer Royal Frank Watson Dyson, and was originally equipped with a 60.9-cm (24-inch) Grubb reflector. It opened in 1929 and in the following years, the university added a donated 20-cm (8-inch) refractor, and another refractor that had been part of the Radcliffe Observatory at Oxford. Today, the observatory continues its educational mission, and has added several new telescopes to its inventory. Five are permanently installed and others are portable. Its main goal is to provide undergraduate students with a chance to do cutting-edge astronomy research on professional-grade equipment.

They also learn computational astronomy and astrophysical theory, all in preparation for further studies if they should choose to take a graduate degree.

The Institute of Astronomy at Cambridge University (IAC) teaches astronomy and fosters research in astrophysics. It maintains a working observatory with a 91-cm (36-inch) telescope. In addition, the Institute has a unique three-mirror telescope, built as a prototype and used for imaging. As part of an agreement made in the late 1980s by what was then the Science and Engineering Council, the United Kingdom paid for a share of the Gemini Observatory. That allowed access by astronomers at IAC to Gemini until the Science and Technology Facilities Council (the successor to SERC) withdrew the UK from the Gemini agreement.

The Royal Observatory at Edinburgh (ROE) in Scotland has always been a part of the University of Edinburgh and its Institute for Astronomy of the School of Physics of Astronomy. First established in the 18th century at Calton Hill, today the observatory is situated at Blackford Hill. The site is owned by the Science and Technology Facilities Council and also contains the UK Astronomy Technology Centre (UK ATC) and a visitor centre. Like its sister institutions at places of higher learning, ROE is involved in astronomy research as well as providing learning space for university classes. Specialised instruments are designed and constructed there, which also provides students and faculty with access to cutting-edge astronomy and astrophysical research equipment.

In addition, ROE was largely responsible for building and installing the UK Schmidt Telescope at the Anglo-Australian Observatory in Coonabarabran, Australia. In Hawaii, the observatory built the UK Infrared Telescope (UKIRT) and the James Clerk Maxwell Telescope. The UK Schmidt Telescope was used, along with plates taken at European Southern Observatory, to make the Southern Sky Survey. It extends the work done by the Palomar Observatory Sky Survey down to far southern latitudes. The UK Astronomy Technology Centre constructed and operated the Visible and Infrared Survey Telescope for Astronomy (VISTA), a 4.1-metre (13.4-foot) infrared-sensitive instrument installed at Paranal Observatory in Chile. It is now part of the European Southern Observatory.

Public Access to the Sky

Observatories hold an important niche in the public mind as places where people can go to look at the sky through a big telescope. In the spirit of outreach, many university, college, and professional observatories offer public access to telescopes, or to tour the places where professional astronomers do their work. Nearly every major observatory has a visitor's centre. Lowell Observatory, for example, has a major public-outreach presence, welcoming tours, offering classes, and access to telescopes and observing areas for amateur skygazers. Observatories also do a lot of 'in person' outreach, presenting talks at star parties, science festivals and meetings, and sending scientists to talk to schools and civic organisations.

There's something very satisfying about visiting an observatory (or its visitor centre) in person. Generally, since these places are built in out-of-the-way places, it takes a bit of advance planning to tour a working facility. On the other hand, some observatories in big cities are relatively easy to visit.

The facility at Royal Observatory Edinburgh, for example, gives tours of its original buildings, instrument collections and books, and public talks about the latest discoveries in astronomy and astrophysics.

Visitors to the Kitt Peak National Observatory start their tour at visitor centre that offers tours, talks, and exhibits.

At Mauna Kea, tourists can stop at the Mauna Kea Visitor Station for refreshments, a gift shop, and talks by naturalists and astronomers. In addition, the centre offers exhibits and special programmes about the local cultural traditions, plus a look at the flora and fauna of the area.

The Very Large Array in New Mexico is well away from the cities, but does attract a lot of tourists who drive out to see radio telescopes spread out across the desert landscape.

Jodrell Bank, home of the Lovell radio telescope in Cheshire, England, has an extensive visitor centre, and offers tours and special events throughout the year. It maintains several pavilions addressing planetary exploration, space missions, and stargazing. There are extensive gardens to visit, as well as a café and gift shop. The centrepiece of the Jodrell Bank experience is, of course, the giant Lovell Telescope.

The European Southern Observatory sites are generally open to visitors on limited occasions, usually on a weekly basis. Given the extreme conditions at the sites, ESO requires pre-registration. At its European headquarters in Garching, Germany, the organisation maintains a full science centre, complete with exhibits and a planetarium. The facility is called the ESO Supernova and it provides public programming as well as programmes for school visits.

Public Observatories

In addition to visitor centre access to observatories, there are a number of public observatories around the world that are specifically aimed at providing people with a chance to look at the sky. One of the most successful of these stand-alone public facilities is the Griffith Observatory in Los Angeles, California. As mentioned earlier in the book, it was constructed as a gift to the people of Los Angeles by wealthy entrepreneur Griffith J. Griffith. Completed in the 1930s, Griffith has welcomed millions of visitors, many of whom have stayed to take a peek at the sky through its telescopes.

In addition to skygazing, Griffith offers extensive exhibits that chronicle humanity's involvement with astronomy. They teach about the history of astronomy and how we came to use the sky, rather than be simply observers of it. The exhibition shows what modern astronomers learn and what has been discovered about the planets, stars, and galaxies.

In a special exhibition, Griffith features the Sun and gives visitors a chance to observe it safely to learn more about its outbursts and activity. One of the most evocative parts of the exhibition is a giant image called 'The Big Picture', which features a part of the sky in the Virgo Cluster of galaxies. This immense image contains stars, asteroids, and galaxies as far as the eye can see. It's intended to give visitors a sense of the immensity of the universe and what astronomers are trying to learn about it. As mentioned earlier, the centrepiece of the exhibitions is the Samuel Oschin Planetarium, where specialised programmes are presented each day. The idea behind the observatory is 'turning visitors into observers', and it gives plenty of opportunity for people to learn more about what it's like to be an astronomer.

A number of science centres around the world also boast observatories, or, at the very least, a collection of telescopes for public use during star party events. At the South Carolina State Museum in Columbia, for example, visitors can see exhibits about astronomy, and then follow up with a chance to observe the night sky in the Boeing Observatory. It features a computer-controlled 1926 Alvan Clark 33-cm (13-inch) refractor, plus a viewing terrace. The facility welcomes school groups, and opens up the observatory for public viewing one night a week. In addition, teachers can attend professional development workshops there.

The Fernbank Science Center near Atlanta boasts the Ralph Buice Jr Observatory, outfitted with a 91-cm, (36-inch) reflector, the largest such telescope in the south-eastern US. It offers free observing sessions twice a week, hosted by an astronomer who gives short talks and guides visitors through the skies.

Not all public observatories are connected to science centres. Some are stand-alone facilities. The Goldendale Observatory in Washington State features both an optical and solar telescope for public viewing, plus a collection of smaller instruments. The observatory, which opened in 1973, has recently been renovated and a set of exhibits has been installed that show the kinds of objects that can be seen through its telescopes. It's owned and operated by Washington State Parks. A ranger offers daily talks and solar viewing sessions, as well as night-time stargazing access. Goldendale came to national attention during 1979 when a solar eclipse swept over the site on 26 February. The US Astronomical League made the observatory its headquarters for the eclipse, which drew thousands of visitors to the windswept site just above the Columbia River Gorge.

In the mid-section of the United States, the Hyde Memorial Observatory in Lincoln, Nebraska, offers access to three telescopes as well as astronomy talks by nationally recognised astronomers and planetary scientists. It was built in 1977 by members of the Prairie Astronomy Club, and was the only observatory at the time built and funded solely by donations from the community. Observatory access is completely free of charge. Hyde's instruments include a 35-cm (14-inch) telescope, a 27-cm (10-inch) instrument, and a 23-cm (9.24-inch) telescope. There's also a solar telescope which lets visitors view the Sun safely by projecting an image of it through the telescope onto a white surface.

In Colorado, the Little Thompson Observatory near Berthoud offers access to the sky for both public and private viewing sessions, and has done so since 1999. It's equipped with several telescopes, including a 45-cm (18-inch) reflector, plus a radio dish. The observatory was first built for the use of students from around the area and is part of Berthoud High School. Nearly 70 per cent of its visitors are part of a school field trip, and more than 75,000 people have passed through the observatory since its opening. One unique offering at Little Thompson is a 'star wall', which shows all the stars in the night sky throughout the year, plus a selection of deep-sky objects, and the constellations of the First Nations tribes, including the Lakota Sioux, who lived in the area more than a century ago.

In the UK, many observatories offer public stargazing access. There is also a small collection of purely public observatories that are not used for research, but are operated by clubs or observing associations. The four most prominent are in Scotland. They include the Airdrie Observatory, curated and operated by the Airdrie Astronomical Association. The facility is open most of the year, with a break for the summer holidays. It caters to families, local schools, and its own club membership. The observatory uses and maintains a 15-cm (6-inch) Victorian refractor under a refurbished dome. In 2010 and 2012, the observatory hosted the Apollo astronauts, Charles Duke, Al Worden, and Richard Gordon, who gave public lectures. The Coats Observatory operates in Paisley, and opened in 1882. It is Scotland's oldest public observatory and is currently undergoing a four-year renovation. There is also Mills Observatory in Dundee, which touts itself as Britain's first purpose-built observatory. It also has a planetarium for public shows about astronomy. One of the latest additions to the public observatory collection is the Scottish Dark Sky Observatory, located on the outskirts of the Galloway Forest Dark Sky Park in southwest Scotland. It lies on the site of the first-ever Dark Sky Park in the UK and Europe, and is operated by the Dark Sky Observatory Group, with partial funding by the government. The facility also has a mobile planetarium and makes its facilities available for public use throughout the year.

As mentioned earlier, the Sydney Observatory offers night-time observatory sessions, day tours, adult astronomy courses, and more. Since light pollution is a problem in the area, only the brightest sky objects – planets, the Moon, and the brighter stars – are observable. During the day, the observatory offers a chance to do solar observing through a special telescope. The regular exhibitions detail the facility's history as a timekeeping station, and its planetarium offers daily star shows.

Online Access to Observatory News

Most of the world's observatories make information available via their websites and through social media. As a good example of an extensive online outreach site, the European Southern Observatory has plenty of information to share. It offers news stories about recent discoveries, in-depth looks at its various telescopes and facilities, and makes a huge array of images available for public use. The Space Telescope Science Institute does the same, sharing an incredible amount of data from the Hubble Space Telescope, along with a library of images, videos, and interviews with its scientists.

Part of this outreach effort is mandated by funding agencies that require taxpayers be informed of how their contributions are being used. This is particularly true in the United States, a practice also employed by other observatories in other countries.

Some observatory websites maintain both public and private entrances – one for the general interest readers and media and the other for the scientists who use the facilities. Others divide their offerings into general and technical pages. Gemini Observatory, for example, has a science dropdown page that guides the reader to information about the telescopes, instruments, technical aspects of planning an observation, and much more. The public pages are more oriented toward news releases about discoveries and so-called 'pretty pictures'.

Observatories also use their websites as a way to share press releases about discoveries being made at their facilities. It's instantaneous communication, which – a century or more ago – would have been made by telegraph or so-called 'snail mail'. Now, within a few seconds, astronomers can share their work around the world with people visiting a facility's online portal. Many observatories also have a large social media presence. This allows them to quickly share information via Facebook, Twitter, YouTube, WeChat, Vimeo, and other channels.

Observatories in the Media and Popular Culture

As useful as observatories are for astronomy, they also attract the attention of filmmakers and other media producers who are quite taken with the dramatic landscapes and 'alien' appearance that some of these facilities have. For example, the venerable Griffith Observatory has appeared in a number of movie and TV programmes. The most famous is the 1955 movie *Rebel Without a Cause*, where James Dean portrays a confused teenager growing up against a gangster backdrop. In one pivotal scene, he has a conversation with a classmate in the planetarium. The observatory commemorates this movie with a bust of James Dean, a popular stopping point for Hollywood tours. Other movies filmed at Griffith include *Dragnet, the Phantom from Space, War of the Colossal Beast, Charlie's Angels: Full Throttle, Transformers, Terminator Salvation,* and *La La Land*. It has also hosted fashion shows, an MTV music awards performance by the group Linkin Park in 2010, and appeared in a number of TV shows, including *The Big Bang Theory, Criminal Minds, MacGyver, Dancing with the Stars*, and made a special cameo appearance in *The Simpsons* as Springfield Observatory.

The radio observatories at Arecibo in Puerto Rico and Socorro, New Mexico, figure prominently in the movie *Contact*, starring Jodie Foster. She plays a radio astronomer whose project for detecting signals from aliens in space is under threat. Her programme is terminated at Arecibo, but she manages to get funding to continue it at the Very Large Array in Socorro. In one scene, widely criticised by radio astronomers, she's shown 'listening' to the signal via headphones. She eventually convinces the powers-that-be she's found a signal coming from outer space, and the movie continues with her being sent her to visit those aliens in a machine that is built using instructions from them.

The Lovell Telescope in England has 'starred' in a number of entertainment pieces. *Dr Who* fans have seen a replica of an earlier version of the telescope used as a stand-in for the shows Pharos Project. It was also used in the *Hitchhiker's Guide to the Galaxy* show, and was part of a story by Fred Hoyle and co-author John Elliot, called *A for Andromeda*. From music videos to other TV shows and postage stamps, the Lovell Telescope is a popular backdrop.

In South America, La Residencia at the Paranal Observatory in Chile was used as a backdrop for the James Bond movie *The Quantum of Solace*. The site stands in for a desert resort where Bond meets with an exiled Bolivian general. The facility, which is located at more than 2,400 metres (7,874 feet) above sea level, provides a comfortable environment in the midst of the desert. Another Bond film, *Goldeneye*, had scenes shot at Arecibo.

In *The Dish*, shot at Parkes Observatory in New South Wales, Australia, filmmakers tell the story of the Apollo 11 landing on the Moon and the role the observatory played in relaying video transmissions to the United States. In reality, Parkes (a radio telescope) was working in tandem with the NASA Honeysuckle Creek Tracking Station near Canberra.

Ground-based observatories aren't the only ones that show up in movies. The Hubble Space Telescope was featured prominently in the beginning of Alfonso Cuarón's movie *Gravity*. During a servicing mission to the telescope, the movie shows how astronauts witness a Russian missile strike on a nearby satellite, the debris from the strike spreads, eventually hitting the shuttle and the telescope, breaking it apart and sending pieces back through Earth's atmosphere.

Hubble itself has been the subject of several documentaries, including the IMAX film *Hubble 3D*. However, in its early 'life', the telescope was a main topic of media interest even before it was launched in 1990. Its original launch date was in October 1986, but the 28 January 1986 explosion of the space shuttle Challenger delayed Hubble's deployment for several years. Once in orbit, scientists quickly discovered the telescope had flawed optics. As a result, the media jumped all over the telescope, its makers, and scientists. The criticism was relentless for several years. In the first year alone, it was the butt of jokes on late-night TV, political cartoons, and even more severe attitudes from the United States Congress and Senate. Hubble was called a 'techno-turkey' and a '1.5 billion-dollar-mistake'. Hubble Space Telescope entered the lexicon as a trope, or a meme, signifying 'failure'. In the *Naked Gun* 2½ movie, the telescope shows up as a picture on a 'wall of shame' showing other failures such as the Edsel car and the *Titanic*. Of course, in the background, during all the public scorn, Hubble scientists were working to fix the problem, and within a few years, the telescope became a symbol of a successful mission. Interestingly, many ground-based telescopes have been using Hubble's sharply produced images as a challenge to do better on their imaging, and some openly brag about getting images 'better than Hubble'.

The movie *Deep Impact* featured an amateur astronomer in an astronomy club studying the night sky. He spots a strange-looking object. This is not so unusual. Amateur astronomers are, on average, far more likely to spot asteroids and comets in the sky since they are more frequently out observing. The interesting trope here, that of the 'lone observer making a discovery', is pretty standard for movie-making, even if it doesn't actually reflect the reality of observing today. Anyway, the student reports the object to the local 'professional' observatory, and this sets off the premise of the movie, which sends a spaceship full of heroes to try to deflect what turns out to be an incoming comet.

The Observatory in Science Fiction

Written science fiction has long featured observatories in stories about major discoveries, aliens, and other adventures among the stars. In his short story *Nightfall* (which was later extended to a novel and made into a movie), Isaac Asimov wrote about astronomers who are working at their observatory on a planet in the middle of a globular cluster. Such a world (if it existed) would have several 'suns' in its sky, and might only experience a full night every few thousand years. The story was inspired by the Ralph Waldo Emerson quote, 'If the stars should appear one night in a thousand years, how would men believe and adore, and preserve for many generations the remembrance of the city of God!' Asimov absolutely believed in the importance of science and clear-minded thinking. Yet, in this story, even the astronomers go mad at the sight of millions of stars which will appear overhead when the last of the planet's six suns sets. The whole event plunges the planet into chaos from which civilization will take thousands of years to recover.

Another classic story, called *The Black Cloud*, written by astronomer Fred Hoyle, posited the discovery of a dark cloud of gas and dust moving toward Earth. Astronomers watch it through telescopes as it approaches. Eventually, it stops and takes up a 'parking orbit' near the Sun. The astronomers come to the conclusion that the cloud is sentient and try to communicate with it.

In the same vein, Arthur C. Clarke's *Rendezvous with Rama* posits a time when asteroids are bombarding Earth, which sets in a motion a Spaceguard-type survey similar to asteroid surveys being done with robotic observatories today. Astronomers at an observatory spy a fairly large asteroid inbound toward Earth and eventually figure out it's actually an alien spacecraft. They nickname it Rama and send astronauts to explore it.

In *Dragonriders of Pern*, author Anne McCaffrey built up a world called Pern, which is periodically endangered by a space-borne biological spore called 'thread'. It appears that this material comes from another nearby world the Pernese call 'The Red Star'. In reality, it's likely a dwarf planet from that system's version of an 'Oort Cloud'. In *Dragonriders*, the civilization on Pern is basically a feudal one, at about the same technical level as Europe was in the 1600s. One of their people invents what appears to be a telescope and uses it in his observatory to chart the motions of the 'Red Star' so as to predict when it might devastate Pern with another threadfall. The rest of the book focuses on how the inhabitants fight thread using fire-breathing dragons and set up a way to scientifically predict the next threadfall by encouraging their scientists to work on the problem.

In many science fiction stories, as well as movies and TV shows, the observatory is almost always presented as a centre of knowledge. Sometimes it's an evil place, or even an alien place where 'egghead' astronomers work well away from the concerns of daily life.

Occasionally, the observatory in a movie or show doesn't even look like our standard idea of what one should resemble. For example, in *Star Trek: The Next Generation*, the ship itself is a travelling observatory. It is equipped with sensor arrays and other equipment that on-board scientists use to study nebulae and other objects the explorers encounter. The Enterprise travels throughout the galaxy, and from time to time, audiences are treated to 24th-century astrophysical explorations built on 20th-century knowledge. For example, in one episode, the ship studied a nebula, travelled into it, and then used its photon torpedoes to illuminate a nebula to look for dark matter.

Probably one of the most provocative stories in science fiction focused on a very different type of telescope in the book *Macroscope,* by Piers Anthony. It takes the idea of a telescope as a machine looking across time and space to a new level. His macroscope is basically a large crystal, sensitive to a newly discovered particle called the 'macron'. By focusing macrons, the macroscope really *can* focus across time and space, and even allow communication between very different times in history. The instrument is installed aboard an orbiting spacecraft that is circling the Sun. There, astronomers with time on it can look at other stars throughout space and time, and try to find worlds around those stars. The larger story involves the discovery of a very highly advanced message, a genius personality buried in the brain of an 'average' person, and a group of people trying to decipher the message. The macroscope observatory serves as a device to look at the human condition and our capability for accepting a future with space travel and other sentient life.

The Observatory as a Meme for Future Exploration

The observatory is, as it has always been, a gateway for travel across time and space. Of course, no one has travelled as far as our eyes (and radio sensors) can 'see'. That, for now, really *is* in the realm of science fiction. However, what the observatory has done for us is unlock possibilities for exploration.

From the first stargazers to the latest Ph.D. student or amateur astronomer who has found something fascinating to study in the sky, the observatory is more than just simply a building or a dish or a detector in a mine shaft. It goes beyond 'place' and 'equipment' and represents an idea, a look into future. That future is filled with answers to questions we haven't yet been able to formulate. It most certainly poses challenges to us and our observatories. Dark matter continues to puzzle theorists, not the least of all because they really can't detect it. All they can find is its effect on the matter that can be detected. Someday an observatory (or, more likely a series of them locked in tandem) will find the evidence we need to show what dark matter really is.

Likewise, at some point in the very near future, a team of astronomers is going to find biotic signatures in the atmosphere of another world. Or, they might find puzzling chemical traces in the ices of Europa or Enceladus, uncovered by spacecraft observatories not yet built.

Somewhere, there's a star just like the Sun starting down the path of old age. Just as likely, there's a group of astronomers who will study the star and chart its actions. What they learn will tell us exactly when our own star will start to die, and what Earth will be like when it does. And, in the not too distant future – perhaps tomorrow – a young lady will look through a telescope in a school observatory and decide, 'This is what I want to do – this is what I want to study'. What will she find?

Look to the observatory. It's the ultimate cosmos mariner – taking our eyes, ears, and minds across the gulfs of space.

APPENDIX

For readers interested in reading more detailed information about specific topics, this appendix serves as a jumping-off point. Observatory equipment is constantly being updated, and most facilities make information about their telescopes and instruments available via their Web pages. The following is a list of sites covering some of the more technical topics.

Adaptive Optics

At least a dozen facilities around the world have adaptive optics systems (or will in the near future). These systems allow astronomers to correct the optics to compensate for turbulence in the atmosphere.

Australian Telescope National Facility https://www.atnf.csiro.au/outreach/education/senior/astrophysics/adaptive_optics.html

Center for Adaptive Optics – Lick Observatory http://cfao.ucolick.org/ao/how.php

European Southern Observatory Adaptive Optics Page https://www.eso.org/public/usa/teles-instr/technology/adaptive_optics/

Gemini Observatory Adaptive Optics (fairly technical) http://www.gemini.edu/sciops/telescopes-and-sites/adaptive-optics

Observatory Costs

Throughout the book, I have included costs for some of the more well-known telescopes and observatories. It's difficult to express a full cost for these places, because they aren't just one-time costs. So, I've given a few, in US dollar amounts, to give rough idea of the kind of money doing astronomy takes. The ongoing costs (for team support, instruments, rentals, purchases, etc.) aren't included. It's probably safe to say that the start-up cost for most observatories are going cost in the millions for a ground-based facility and the billions for space-based telescopes and probes.

Observatory Instruments

Every observatory maintains a suite of instruments, which can be attached to the telescope as needed. Generally, these include cameras and imagers, spectrographs, and photometers. Space telescopes are launched with instruments as well, which serve throughout the life of the spacecraft. The only telescope currently that can be serviced is the Hubble Space Telescope, and it has had several of its instrument 'swapped out' during various servicing missions.

Gemini Observatory Instruments and Science Operations
https://www.gemini.edu/sciops/instruments/

Hubble Space Telescope Instruments and Servicing History
https://www.nasa.gov/content/goddard/hubble-space-telescope-science-instruments

National Optical Astronomy Observatory
This site has an extensive list of observatory instruments with links to technical descriptions.
http://ast.noao.edu/observing/current-telescopes-instruments

Very Large Telescope Instruments (Chile)
https://www.eso.org/public/teles-instr/paranal-observatory/vlt/vlt-instr/

W.M. Keck Observatory Instruments Suite
https://www2.keck.hawaii.edu/inst/index.php

Telescope Mirrors
Making a Mirror for a Large Telescope
One of the best descriptions of making a telescope mirror was reprinted by *Scientific American* under a Creative Commons license at: https://www.scientificamerican.com/article/how-to-build-an-80-foot-wide-telescope-mirror-to-see-deep-space/

This article describes the process by which the Giant Magellan Telescope mirror was made.

Timekeeping and Calendar-making
As mentioned earlier, timekeeping and calendar-making depended on knowledge of the sky. The Sun was the ancients' primary way of telling time, and of course, they noticed the Moon cycled through its phases in 27 days. The next step was to

take that knowledge and devise a yearly calendar. Most ancient civilizations did this fairly early on, and kept adjusting their calendars to account for the actual lengths of day, month, and year. Today, the Gregorian calendar (developed in 1582) is widely in use, although people also use others, including the Chinese, Buddhist, Hebrew, Japanese, the Islamic calendar, and so on.

Throughout history, determining 'local' time and longitude were essential for navigation. The great observatories at Greenwich, Armagh, Paris, Washington DC, and others were initially established to determine local time and chart the positions of stars so that mariners had accurate tables on which to base their travels. Traveling across longitudes required very accurate timekeeping. Sailors could determine their latitude by knowing where they were on Earth relative to the position of a fixed star in space. But, determining longitude was more difficult and longitude is related to local time. So, accurate timekeeping was important. It wasn't until the 18th century that a chronometer was made that would keep time at sea, although it was far from a perfect solution. Another way to figure out longitude was to use lunar distance as a way to calculate longitude. The development of the telegraph in the 1830s provided a new way to send time signals to whoever needed them. In 1865, the US Naval Observatory began sending out time signals via telegraph, which helped to synchronise time not only for mariners but for astronomers needing accurate signals when observing events such as eclipses.

In the 20th century, time signals were broadcast via wireless telegraphy from the Eiffel Tower in Paris, and from Washington, and London. Using those signals, mariners were able to correct their on-board chronometers. Today, positional information is not so dependent on lunar measurements or other methods. Instead, GPS signals contain time code that allows people around the world to maintain accurate local time. This constellation of satellites is one of many methods people use for timekeeping and positional information.

BIBLIOGRAPHY

Books and Articles

Duffner, Robert W., and Robert Q. Fugate. *The Adaptive Optics Revolution: A History*, University of New Mexico Press, 2009.

Clements, D.L. *Infrared Astronomy – Seeing the Heat*, CRC Press, 2014.

David, L. *Mars: Our Future on the Red Planet*, National Geographic Books, 2016.

Dick, S.J. *Sky and Ocean Joined: The U.S. Naval Observatory 1830–2000*, Cambridge, UK, Cambridge University Press, 2002.

Fabian, A.C., Ed. *Frontiers of X-ray Astronomy*, Cambridge University Press, 2012.

Hoskin, M. *The Cambridge Illustrated History of Astronomy*, Cambridge University Press, 2017.

Ingalls, Albert G. *Amateur Telescope Making*, Willmann-Bell, 1996.

Kloeppel, J.E. *Realm of the Long Eyes: A Brief History of Kitt Peak National Observatory*, Univelt Publishers, 1983.

Krupp, E.C. *Echoes of the Ancient Skies*, Dover Books, 2012.

Krupp, E.C. *Skywatchers, Shamans, and Kings*, Wiley, 2012.

Leverington, D. *Observatories and Telescopes of Modern Times: Ground-based Optical and Radio Astronomy Facilities since 1945*, Cambridge University Press, 2017.

Lockman, J.F., Ghigo, F.D, Balser, D.S. Eds, *But It Was Fun: the First Forty Years of Astronomy at Greenbank*, NRAO.

Petersen, C.C. *Astronomy 101: From the Sun and Moon to Wormholes and Warp Drive, Key Theories, Discoveries, and Facts about the Universe*, Adams Books, 2013.

Petersen, C.C. 'High Stakes Astronomy at Low Frequencies', *State of the Universe: New Images, Discoveries, and Events*, Springer Praxis Books, 2008.

Petersen, C.C. Brandt, J.C. *Hubble Vision,* Cambridge University Press, 1998.

Petersen, C.C., Brandt, J.C. *Visions of the Cosmos,* Cambridge University Press, 2003.

Rawlins, D. 'Great Pyramid Alignment'. *Griffith Observer,* May 2019.

Renshaw, S., Ihara, S. 'A Brief History of Amateur Astronomy in Japan', *Sky & Telescope* Magazine, March 1997.

Scherrer, D. Ancient Observatories—Timeless Knowledge, Stanford Solar Center http://solar-center.stanford.edu/AO/Ancient-Observatories.pdf

Smith, R.W. 'The Observatory', *A Companion to the History of Science,* Wiley Books, 2016.

Sobel, Dava. *Longitude: The True Story of a Lone Genius who Solved the Greatest Scientific Problem of His Time,* Bloomsbury, 2010.

Stern, A., Grinspoon, D. *Chasing New Horizons: Inside the Epic First Mission to Pluto,* Picador Press, 2018.

Tyson, N.D. *Astrophysics for People in a Hurry,* W.W. Norton & Company, 2017.

Tucker, W. H. *Chandra's Cosmos: Dark Matter, Black Holes, and Other Wonders Revealed by NASA's Premier X-Ray Observatory,* Smithsonian Books, 2017.

Selected Educational Websites

Astronomical Features in Churches
https://www.atlasobscura.com/articles/catholics-built-secret-astronomical-features-into-churches-to-help-save-souls

A History of Scientific Cosmology
https://history.aip.org/history/exhibits/cosmology/index.htm

Blast!
http://www.blastthemovie.com/homepage/

The Dark Energy Survey
https://www.darkenergysurvey.org/

The Electromagnetic Spectrum
https://science.nasa.gov/ems/

History of Gamma-ray Astronomy
http://teacherlink.ed.usu.edu/tlnasa/reference/imaginedvd/files/imagine/docs/science/know_l1/history_gamma.html

History of Islamic Spain
http://www.sjsu.edu/people/patricia.backer/history/islam.htm

Bibliography

History of Stonehenge
https://www.english-heritage.org.uk/visit/places/stonehenge/history-and-stories/history/

History of Observatories
https://amazing-space.stsci.edu/resources/explorations/groundup/

Introduction to Adaptive Optics and Its History
http://www.cfao.ucolick.org/EO/Resources/History_AO_Max.pdf

Large-Scale Structure in the Universe
http://astronomy.swin.edu.au/cosmos/L/Large-scale+Structure

The First Stars in the Universe
https://www.scientificamerican.com/article/the-first-stars-in-the-un/

The Starry Messenger (Cambridge)
http://www.sites.hps.cam.ac.uk/starry/starrymessenger.html

Ultraviolet Astronomy
https://www.esa.int/Our_Activities/Space_Science/Observations_Seeing_in_ultraviolet_wavelengths

What is Radio Astronomy?
https://public.nrao.edu/radio-astronomy/what-is-radio-astronomy/

Why Do We Observe Gamma rays?
https://www.esa.int/Our_Activities/Space_Science/Integral/Why_do_we_observe_gamma_rays

Selected Observatory and Project Websites
Airdree Public Observatory
https://www.airdrieobservatory.com/

Akatsuki Mission
https://solarsystem.nasa.gov/missions/akatsuki/in-depth/

Anglo-Australian Observatory
https://www.aao.gov.au/

Ariel Space Mission
https://arielmission.space/

Armagh Observatory

https://www.armagh.ac.uk/shorthistory.php

Birmingham Solar-Oscillations Network (BiSON)
http://bison.ph.bham.ac.uk/

CalTech Optical Observatories
http://www.astro.caltech.edu/palomar/homepage.html

Chandra X-Ray Observatory
http://chandra.harvard.edu/

CSIRO – Australian Astronomy and Space Site
https://www.csiro.au/en/Research/Astronomy

Dunsink Observatory
https://www.dunsink.dias.ie/

Deep Underground Neutrino Experiment
https://www.dunescience.org/

European Southern Observatory
https://www.eso.org/

Fermi Gamma-ray Space Telescope
https://fermi.gsfc.nasa.gov/

Gemini Observatory
https://www.gemini.edu/

Giant Magellan Telescope Book
https://www.gmto.org/gallery/gmt-resources/#GMT_Science_Book_2018

HabEx Mission
https://www.jpl.nasa.gov/habex/

Harvard College Observatory
https://www.cfa.harvard.edu/hco

Hubble Space Telescope
http://hubblesite.org/

James Webb Space Telescope
https://webbtelescope.org/webb-science

Kepler Telescope
https://www.jpl.nasa.gov/missions/kepler/

Kitt Peak National Observatory
https://www.noao.edu/kpno/

KM3NeT
https://www.km3net.org/

Leibniz Institute for Astrophysics-Potsdam
https://www.aip.de/en/institute

LIGO Observatory
https://www.ligo.caltech.edu/

Los Cumbres Observatory
https://lco.global/

Mount Wilson Observatory
https://www.mtwilson.edu/

Murchison Widefield Array
http://www.mwatelescope.org/

National Radio Astronomy Observatory
https://public.nrao.edu/

NuSTAR Mission
https://www.nustar.caltech.edu/

Pierre Auger Observatory
https://www.auger.org/

Royal Observatory Edinburgh
https://www.roe.ac.uk/

Royal Observatory Greenwich
https://www.rmg.co.uk/royal-observatory

Slooh Observatory Network
https://slooh.com/

SOHO Solar Mission
https://sohowww.nascom.nasa.gov/

Solar Dynamics Observatory (SDO)
https://sdo.gsfc.nasa.gov/

South African Astronomical Observatory
https://www.saao.ac.za/

Spitzer Space Telescope
http://www.spitzer.caltech.edu/

Square Kilometre Array (SKA)
https://www.skatelescope.org/

Subaru Telescope
https://subarutelescope.org/

Super-Kamiokande Neutrino Detector
http://www-sk.icrr.u-tokyo.ac.jp/sk/index-e.html

The CHARA Array
http://www.chara.gsu.edu/public/instrumentation/31-the-chara-array

Thirty Metre Telescope
https://www.tmt.org/

United States Naval Observatory
https://www.usno.navy.mil

Virtual Telescope Project
https://www.virtualtelescope.eu/

WFIRST Observatory
https://wfirst.gsfc.nasa.gov/

XCaliber Project
https://source.wustl.edu/2018/12/second-scientific-balloon-launches-from-
    antarctica/

XMM-Newton
https://www.cosmos.esa.int/web/xmm-newton